Fishes of the Okefenokee Swamp

Fishes of the
Okefenokee Swamp

Joshua Laerm and B. J. Freeman

The University of Georgia Press

Athens and London

Reissued 2007
© 1986 by the University of Georgia Press
Athens, Georgia 30602
www.ugapress.org

Designed by Kathi L. Dailey
Set in 10 on 13 Linotron 202 Ehrhardt
Typeset by The Composing Room of Michigan
Printed digitally in the United States of America

The Library of Congress has cataloged the hardcover
edition of this book as follows:
Library of Congress Cataloging-in-Publication Data

Laerm, Joshua.
Fishes of the Okefenokee Swamp / Joshua Learm and
B.J. Freeman.
ix, 118 p. : ill. ; 24 cm.
Bibliography: p. 111-116.
Includes index.

ISBN 0-8203-0820-X (alk. paper) — 0-8203-0841-2
(pbk. : alk. paper)
1. Freshwater fishes—Okefenokee Swamp
(Ga. and Fla.)—Identification.
2. Fishes—Okefenokee Swamp
(Ga. and Fla.)—Identification. I. Freeman, B. J.
II. Title.
QL628.O36 L34   1986
597.092'975875 19   85-16486

Paperback reissue ISBN-13: 978-0-8203-3135-5
ISBN-10: 0-8203-3135-X

British Library Cataloging-in-Publication Data
available

For
Donald C. Scott
teacher and friend

Contents

Acknowledgments   ix

Introduction   1

   What's About a Fish?   4
   The Okefenokee Swamp   9
   History of Ichthyology in the Okefenokee Swamp   16
   Comparison with Regional Fauna   18
   How to Use the Keys   19

Key to the Families   21

Fish Families and Keys to the Species   23

List of the Species   31

Fishes of the Okefenokee Swamp   33

   Florida Gar   34
   Bowfin   36
   American Eel   38
   Redfin Pickerel   40
   Chain Pickerel   42
   Eastern Mudminnow   44
   Lake Chubsucker   46
   Spotted Sucker   48
   Yellow Bullhead   50
   Brown Bullhead   52
   Channel Catfish   54
   Tadpole Madtom   56
   Speckled Madtom   58
   Pirate Perch   60

Golden Topminnow   62

Banded Topminnow   64

Lined Topminnow   66

Pygmy Killifish   68

Mosquitofish   70

Least Killifish   72

Brook Silverside   74

Everglades Pygmy Sunfish   76

Okefenokee Pygmy Sunfish   78

Mud Sunfish   80

Flier   82

Blackbanded Sunfish   84

Bluespotted Sunfish   86

Banded Sunfish   88

Warmouth   90

Bluegill   92

Dollar Sunfish   94

Spotted Sunfish   96

Largemouth Bass   98

Black Crappie   100

Swamp Darter   102

Blackbanded Darter   104

Glossary   107

References   111

Index   117

Acknowledgments

A number of individuals have contributed measurably to this book. John Rappole, Laurie Vitt, Liz McGhee, Gene Helfman, and Lloyd Logan provided valuable field assistance.

The illustrations, drawn from original specimens, were prepared by Monica Springer, Debbie Dubac, and David Fisher as senior theses in the Scientific Illustration Interdisciplinary Studies Program at the University of Georgia under the supervision of Lloyd Logan, who also provided several illustrations.

Liz McGhee, Donald Scott, Gene Helfman, Gary Grossman, and Mary Freeman read the manuscript and provided many helpful comments.

We extend our thanks to Debbie Winter of the University of Georgia Press for the high quality of her editorial suggestions as well as to Karen Orchard for her continued interest.

Support for this project was provided by the University of Georgia Museum of Natural History and an N.S.F. grant (BSR 7808842) to the University of Georgia Research Foundation.

This is a contribution of the University of Georgia Museum of Natural History.

Introduction

This book is about the fishes that occur in the Okefenokee Swamp and its surrounding watershed and is intended to serve as an identification manual. While it is written primarily for the general public with perhaps limited formal training in fish biology, we have also provided information that may be useful to professional biologists.

As educators and professional biologists who deal extensively with the public, we have found that persons with more than just a casual interest in fish (such as fishermen or naturalists) are interested in certain types of information. Thus we have included what we think is useful and informative to a general readership. We believe good, high quality illustrations depicting each species are essential. Similarly, sufficient verbal information to permit identification of each species is also important. In our experience, nonbiologists are interested in general aspects of natural history and biology, not extensive data, tables, and details of taxonomy. For those who are interested in further information and access to this literature, we provide references. These are given numerically within the text and cited fully in the list of references at the end of the book.

Arrangement of the Taxa

The arrangement of the various species described in this book follows accepted taxonomic practices used by ichthyologists (fish specialists). Thus, *orders* are arranged in a phylogenetic (evolutionary) sequence with more primitive groups (like the gar or bowfin) listed first and more advanced groups (like the perches) listed last. Within *orders*, the constituent *families* are also listed phylogenetically. The various *genera* and *species* are listed alphabetically.

Common Names

Since the local common name of a species may frequently differ throughout its range, the use of the scientific name is often preferred by professional biolo-

1

gists. The common names of the fishes of the Okefenokee Swamp used here are those recognized by the American Fisheries Society (75). We have listed other local common names insofar as we are aware of them.

The Species Concept

Basically, a species is made up of a complex of interbreeding (or potentially interbreeding) individuals that share a common inheritance. This infers that one particular species is reproductively isolated or separate from other species. In general, a species is recognizable as a distinct "type" of organism, although this typological approach is not always a satisfactory criterion. Essentially, individuals of separate species do not normally interbreed. In some cases, however, interbreeding may occur, resulting in hybrids (a condition commonly occurring in sunfishes). The hybrids themselves are usually sterile or their offspring not viable. As a result, species tend to retain their distinction from other species.

Taxonomic Nomenclature

Each species is recognized by a unique scientific name consisting of two Latinized words. The first word is the generic name, the first letter of which is always capitalized. The second word is the species name. Some species may have recognizable subspecies. These are usually geographic variants distinguishable by fine details of their morphology or behavior. In this case there may be three Latinized words, the third referring to the subspecies. With a few exceptions we have not distinguished subspecies in this book.

The species name is normally followed with the name of the person who originally described it, as we have done at the beginning of each species account. If the person's name appears in parentheses, this indicates that the generic name has been changed after it was originally described. This commonly happens as ichthyologists learn more about the interrelationships of the various groups.

Consider as an example the largemouth bass, *Micropterus salmoides* (Lacepède). *Micropterus* is the generic name of most of our freshwater bass, and *salmoides* is the species name for the largemouth bass. The name *Micropterus* is derived from Greek meaning "small fin" and *salmoides* from the Greek word *salmo* meaning "trout," as this fish is often referred to as a "trout" in the South. Lacepède originally described the species in 1802 from specimens taken in South Carolina. His name is in parentheses because the species was removed

2

from the genus to which it was originally assigned (*Labrus*) and referred to another (*Micropterus*).

If we were to concern ourselves with geographic variation in the region, it might be necessary to discuss various subspecies. For example, *Micropterus salmoides salmoides* is the subspecies of largemouth bass occurring in north-western Georgia. However, populations of largemouth bass throughout most of the southern part of peninsular Florida are somewhat different in their markings. They are referred to as a distinct subspecies, *Micropterus salmoides floridanus*. The problem of distinct subspecies can be very complex, as remaining populations of largemouth bass throughout Georgia are considered to be intergrades between the two subspecies discussed above (3).

Organization of the Accounts

For each species occurring in the Okefenokee Swamp and its watershed, we provide the scientific name and standard common name as well as an illustration and other local names. The derivations of the scientific names are also included. In many cases these are very descriptive and helpful in our understanding as to how a particular species is named.

The *description* is intended to readily provide sufficient information to enable the reader to distinguish one species from another. There are usually two paragraphs for each species description; the first describes general characteristics, the second coloration and markings.

The *similar species* section provides a brief comparison of other species with which it might be confused.

The *distribution* section describes the natural range of the species in North America.

The *habitat* section gives the habitat preferences of the species throughout its range as well as specific habitat distribution in the swamp.

In the *biology* section the discussion is limited to feeding and breeding aspects of the species' life history. We provide what information is known about these fishes in the Okefenokee Swamp or adjacent regions in the Southeast. When no regional information is available, we have relied on published information from elsewhere throughout the species' range.

The *comments* section deals with such topics as the species' relationship to man as a food or game fish or other aspects of its biology that may be of interest to the reader. Discussion of subspecies or other taxonomic questions of interest to professional biologists are also presented in this section.

What's About a Fish?

Although most everyone knows a fish when they see one, it is very difficult for even a trained biologist to accurately define the term *fish*. The problem is that fishes as a group are extremely diverse. There is a popular misconception that vertebrate animals (those with a backbone) are divided into 5 classes: fishes, amphibians, reptiles, birds, and mammals. This is not entirely true. In fact, there are (depending on the authority one consults) 6–7 separate classes of vertebrate animals that are considered to be fishes. This makes fishes more diverse as a group than all other vertebrates combined. Thus the term *fishes* does not really reflect a natural group of organisms as do the terms *reptiles* or *mammals*. While this may seem to be a trivial point best left to the ponderings of specialists, it is nonetheless important. Fortunately, almost all living fishes belong to one of two classes of *fishes*. These are the chondrichthyan (or cartilaginous) fishes, which include the sharks, skates, and rays, and the osteichthyan (or bony) fishes. The osteichthyan fishes include almost all the remaining forms of fishes with which the reader is familiar. These are the fishes we catch, the fishes we eat, and the fishes we enjoy as pets.

The use of the terms *fish* and *fishes* requires some comment. Generally, *fish* is used to describe an individual specimen or a distinct species, while *fishes* refers to numbers of individuals or groups of species.

The fishes that occur in the Okefenokee Swamp are osteichthyan fishes and share a number of morphological (structural) features. When we think of a "fish," our image is most probably that of an osteichthyan fish. Those features that bony fishes share, however, are in large part unique to the group. Other classes of fishes do not necessarily possess the same features. For example, lampreys are fishes, yet they lack scales, paired fins, jaws, and even bony vertebrae. These fishes are very primitive vertebrates and are included in an entirely different class from those with which we are concerned in this book. With this in mind perhaps we can now get to the point—what's about a fish, an osteichthyan fish!

To use this book for identifying fishes, it is necessary to have some knowledge of fish anatomy, particularly those features that are used to distinguish one species from another. Even the casual reader who does not plan to use this book as an identification manual should have some interest in anatomy because almost everything about a fish reflects its adaptation to an aquatic environment.

The reader first should be familiar with a number of general anatomical dimensions. *Dorsal* refers to the top or the back, whereas *ventral* refers to the bottom or belly. *Anterior*, of course, refers to the front, and *posterior* to the hind or tail end. *Lateral* refers to the sides or toward the sides of the body, while *medial* refers to the midline or toward the midline of the body.

Most fishes are typically elongate or fusiform, shaped something like a torpedo. There are exceptions, of course—like the eel, which is extremely elongate, almost serpentine. In general, the body of a fish may be divided into three parts: the head, the trunk, and the tail.

The *head* of a fish extends from the tip of the snout to the posterior margin of the *operculum*, a bony flap covering the *gills*.

The position of the *mouth* is highly variable in fishes, depending on their mode of feeding. The mouth is said to be *terminal* when it forms the front of the head, *subterminal* when the snout extends beyond the mouth as in the suckers, or *superior* when the lower jaw extends beyond the upper jaw as in the topminnows. The mouth is made up of upper and lower jaws. The lower or mandible jaw is composed of a series of paired bones on either side, the most prominent being the *dentaries*, which usually bear teeth. The upper jaw is also composed of a series of paired bones. The anteriormost are the *premaxillaries*. These may be separated from the remainder of the snout by a distinct groove in fishes that have *protractile* premaxillary bones. In fishes without protractile premaxillaries these

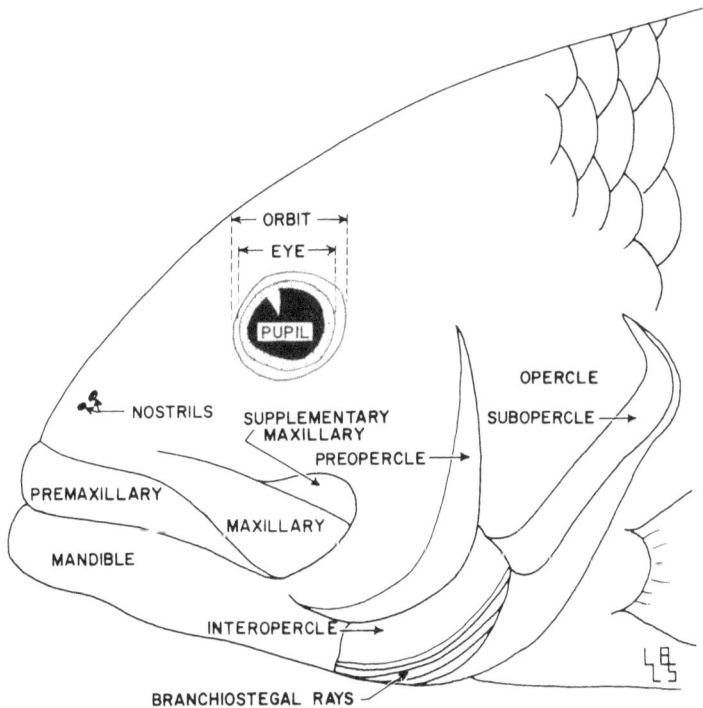

bones are directly connected to the snout by a fleshy bridge, the *frenum*. The *maxillary* bones lie above and behind the premaxillary bones. The posterior border of the maxillary marks the posterior end of the mouth.

On either side of the midline of the snout in front of the eyes are two pairs of *nostrils*. The two pits of each pair are separated by a flap of skin. These pits are sensory in function. The eyes are located in a cavity called the *orbit*.

A series of bones covers the gills. These make up the so-called *operculum* and include the large *opercle*. Posterior and ventral to the *opercle* is the *subopercle*. Ventral to this and somewhat anterior to it is the *interopercle*. A *preopercle* bone lies anterior to the opercle and dorsal to the interopercle. By lifting the posterior margin of the operculum, the gills may be seen. This is where the exchange of respiratory gases occurs.

The *teeth* in fishes are found on a number of bones in the upper jaw. Teeth are usually present on the premaxillary. In the middle of the roof of the mouth is the unpaired *vomer* bone, which may bear *vomerine* teeth. Lateral to the vomer are *palatine* bones, which may bear *palatine* teeth. In the lower jaw the dentary bones typically bear teeth. The *tongue* (glossus) may also bear teeth.

Some fishes bear accessory filamentous sensory structures on the head called *barbels*. These are found most typically in the catfish.

Ventral to the operculum on either side is a series of *branchiostegal rays* that support the gill membranes. The fleshy region of the throat between the branchiostegal rays is the *isthmus*. Some primitive fish like the bowfin possess a distinct bony *gular plate* lying between the anterior part of the right and left lower jaws.

The *trunk* extends from the operculum to the *anus* or *vent*. Located in the trunk are pectoral and pelvic fins as well as the dorsal fin. The *thorax* or *breast* is the ventral part of the trunk in front of the pectoral fins; the *abdomen* or *belly* is the ventral part of the trunk from the pectoral fins to the anus.

The *tail* extends from the anus to the end of the caudal fin. It includes the region in which are located the anal fin and, when present, the adipose fin. The *caudal peduncle* is the constricted portion of the tail.

Most fishes possess a series of *appendages* or *fins*—the paired lateral fins and unpaired median fins. The *pectoral fins* are the frontmost of the paired fins and are located directly behind the head. They are usually in front of and above the paired pelvic fins. The *pelvic fins* are also paired and are usually located mid-trunk. In some species, however, they may be located far forward, directly behind and below the pectoral fins.

The unpaired median fins may be of some help in distinguishing closely related species. The *dorsal fin* is located in the middle of the back. The elements

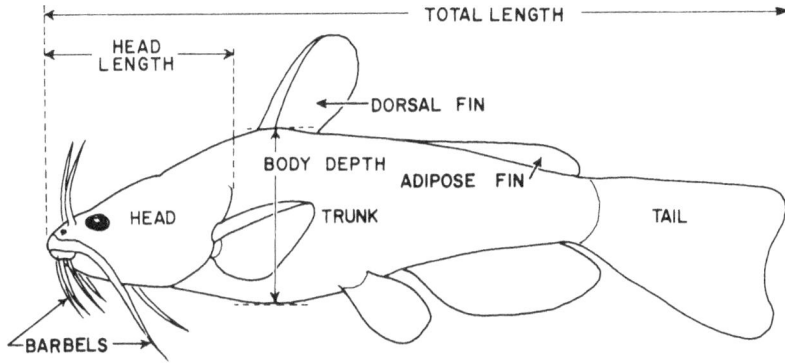

Measurements of a Fish

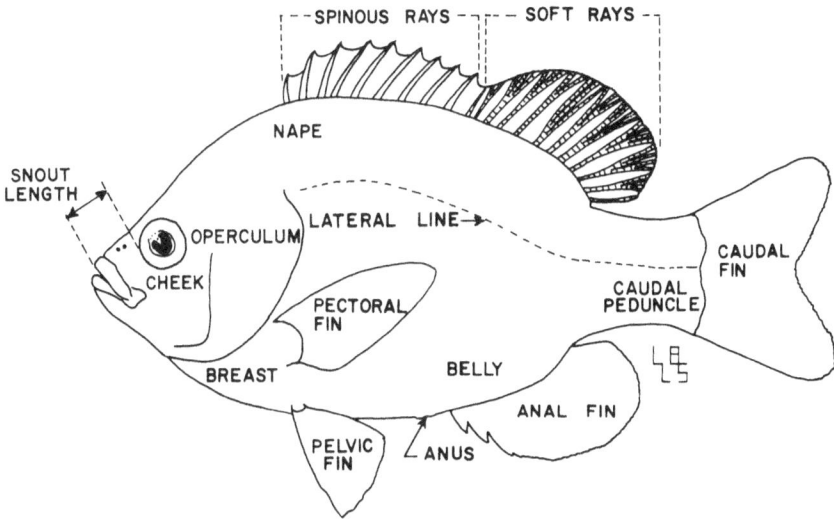

External Anatomy of a Fish

that support the membrane are *fin rays*. These may be soft or spinous and are easily distinguished. Persons who have handled a catfish or a bream may recall their first painful exposure to the *spinous rays*, which are hard and sharply pointed. Spinous rays are not branched and are not segmented. *Soft rays* are soft, segmented, and quite often branched when viewed from the side. There may be a single dorsal fin without spines, as in the suckers, or a single dorsal fin with both spinous and soft rays, as in bream. Some species, like the darters, have two separate dorsal fins. The *anal fin* is located in the midventral region behind the anus. Usually there is not more than one anal fin, which may have spinous and soft rays. It is important to be able to distinguish between soft and spinous rays because the total number of rays, as well as the relative number of soft and spinous rays, are so important in fish classification.

The *caudal fin* is supported by soft rays. Considerable variation occurs in the shape of this fin. Primitive bony fish (and most sharks) have an *heterocercal tail*, in which the axis of the vertebral column extends out into the upper lobe of the fin. A modified form of this, the *abbreviate heterocercal tail*, occurs in the bowfin, *Amia*, and the gar, *Lepisosteus*. However, very young specimens of these fishes have heterocercal tails. Most extant bony fishes have a *homocercal* tail. In this tail the vertebral column ends at the base of the fin, and the fin itself is usually symmetrical top and bottom. Furthermore, this fin may be deeply forked (concave), rounded (convex), even pointed, or almost any shape between these extremes.

An *adipose fin* is present in catfish. This is an unpaired median fin located behind the dorsal fin. It is a fleshy structure and has no fin rays.

The most characteristic features of bony fishes are the *scales*. Some fishes, however, like catfish, lack scales. Even fishes with scales may have areas where the scales are absent. These areas are said to be *naked*. There are several types of scales. Primitive fishes like the gar have *ganoid* scales. These are thick bony scales that do not overlap and are rhomboid or diamond-shaped. Higher bony fishes possess thin, overlapping bony-ridge scales that are typically thin and translucent. These are formed by concentric layers of bone that are laid down as the fish grows. This growth produces a series of *annulae* that can be used to age a fish. A series of ridges called *radii* emanate outward from the center of the scale. The most simple of the bony-ridge scales is the *cycloid* scale. Very similar to this is the *ctenoid*, which is characterized by tiny spines covering the exposed posterior portion of the scale.

Most bony fishes possess a complex *lateral line system.* This is made up of a

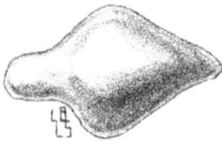

GANOID SCALE   CYCLOID SCALE   CTENOID SCALE

8

series of pores located in the scales that connect with an underlying canal. The lateral line system is sensory in function, receiving pressure vibrations in the water. The most obvious part of this lateral line system is the *lateral line* that may be observed running horizontally along the sides of the trunk from behind the gills to the tail. The number of scales in the lateral line is often used in classification of various species and subspecies of fishes.

A number of measurements are frequently made of fishes to aid in classification. The most important of these include: *total length, body depth, head length, snout length,* and *scale count.* These measurements are described in the glossary.

The Okefenokee Swamp

Hydrology

The Okefenokee Swamp, located in southeastern Georgia and northern Florida, is the largest freshwater wetland in the United States. The swamp proper (1,775 square kilometers) and its watershed cover an area of some 3,800 square kilometers—about the size of the state of Rhode Island. The watershed, which lies mainly to the northwest, consists of the uplands surrounding the swamp. About four-fifths of the swamp and a small portion of the watershed is included in the Okefenokee National Wildlife Refuge, administered by the U.S. Fish and Wildlife Service.

Water levels in the swamp are controlled by the interaction between water input and output and evapotranspiration (the amount of water lost by evaporation). Water input to the swamp occurs through precipitation. Rainfall and stream runoff from the large northwestern watershed contribute to the Suwannee River drainage, while the eastern portion of the swamp (St. Marys River drainage) receives most of its input chiefly from rainfall. It is believed that deep groundwater has a negligible effect on water balance in the swamp, which is perched above the regional water table. Reports that springs feed the swamp have not been substantiated, although recent geological evidence suggests that some surface springs may exist (76).

The Okefenokee is a relatively shallow-water swamp. Under normal hydrological conditions the peat soil is inundated by water. Although certain areas of the swamp may have water depths of more than 8 feet, the average is about 2 feet. Of course, during drought conditions water levels may be significantly lower to nonexistent in certain areas. There has, however, been an increase in

The Okefenokee Swamp

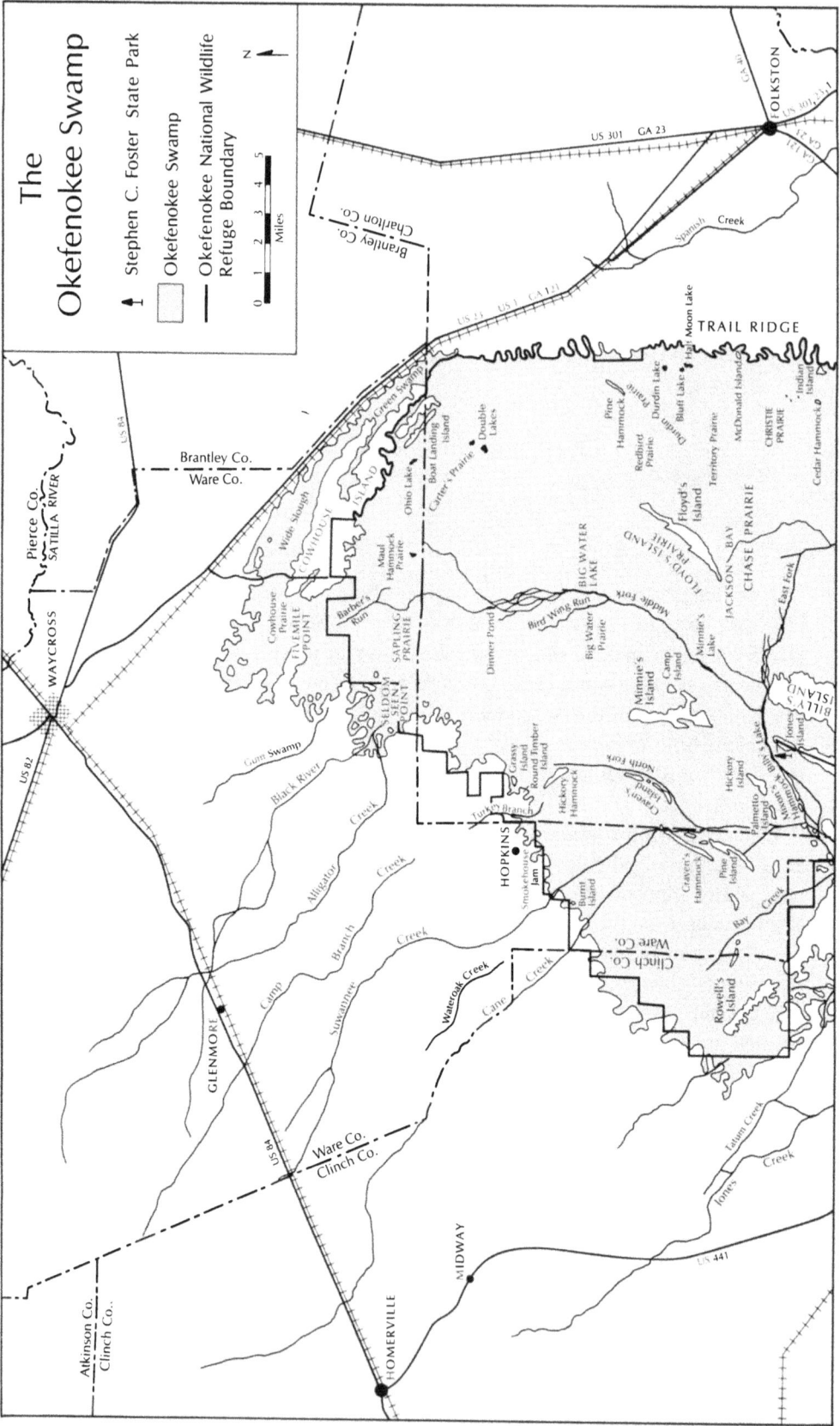

N →

Miles
0 1 2 3 4 5

FOLKSTON

US 301 GA 23

Spanish Creek

TRAIL RIDGE

High Moon Lake

Indian Island

Cedar Hammock

CHRISTE PRAIRIE

McDonald Island

Bluff Lake

Durdin Lake

Territory Prairie

Pine Hammock

Redbird Prairie

Durdin Prairie

FLOYD'S ISLAND

Floyd's Island

CHASE PRAIRIE

JACKSON BAY

US 23 US 1 GA 121

Green Swamp

COWHOUSE ISLAND

Brantley Co.
Charlton Co.

Brantley Co.
Ware Co.

Pierce Co.
SATILLA RIVER

US 84

Wdr. Slough

Ohio Lake

Boat Landing

Double Lakes

Carter Prairie

Maul Hammock Prairie

SAPLING PRAIRIE

Barber's Run

TEN MILE POINT

Cowhouse Prairie

SELDOM SEEN POINT

Big Water Prairie

BIG WATER LAKE

Bird Wing Run

Dinner Pond

Middle Fork

Minnie's Lake

Camp Island

Minnie's Island

BILLY'S ISLAND

East Fork

WAYCROSS

US 82

Gum Swamp

Black River

Alligator Creek

Grassy Island
Round Timber Island

Hickory Hammock

Turkey Branch

North Fork

Craven's Island

Hickory Island

Jones' Island

Billy's Lake

Palmetto Hammock

Mixon's

Craven's Hammock

Pine Island

Smokehouse Lake

HOPKINS

Burnt Island

Bay

Rowell's Island

Jones Creek

Clinch Co.
Ware Co.

Camp Creek

Suwannee Creek

Branch

Watercak Creek

Cane Creek

Creek

Ware Co.
Clinch Co.

US 84

GLENMORE

MIDWAY

US 441

Tatum Creek

Jones Creek

Atkinson Co.
Clinch Co.

HOMERVILLE

FOLKSTON

US 1 US 23 US 301

US 301 GA 23

FLA 127

RIVER

SAINT GEORGE

FLORIDA
GEORGIA

TOLEDO   ST. MARYS

GA 23 GA 121

US 1 GA 121

GA 94

MONIAC

TRAIL RIDGE   CAMP CORNELIA

Pine Hammock
Durdin Lake
Bluff Lake
Half Moon Lake
Indian Island
McDonald Island
Redbird Prairie
Territory Prairie
Floyd's Island
CHRISTIE PRAIRIE
Cedar Hammock
Duck Island
Mizell Prairie
Cooter Lake
Chesser Prairie
Chesser Island
Seagrove Lake
Trout Lake
Trout Lake Prairie
Horse Island
Monkey Lake
Coward Lake
Soldiers' Camp Island
Buck Prairie
Hog Island
Grand Prairie
Buzzard's Roost Lake
Number One Island
Cane Pond
Carnet Lake

BIG WATER LAKE
Big Water Prairie
JACKSON BAY
FLOYD'S ISLAND PRAIRIE
LLOYD'S ISLAND
CHASE PRAIRIE
DOUBLE O BAY
Suwannee Canal
East Fork
Middle Fork
Charlton Co.
Ware Co.

Minnie's Lake
Camp Island
Minnie's Island
BILLY'S ISLAND
Honey Island
Mitchell Island
Broomstraw Island
MOONSHINE RIDGE
South Prong
Moccasin Creek
MOCCASIN SWAMP
Mim's Island

Hickory Island
Craven's Island
Palmetto Island
Pine Island
Jones Island
THE POCKET
Billy's Lake
Billy's Lake
Honey Island Prairie
Bugaboo Island
Floyd Island
Honey Island
Honey Scrub Island
Jack Island
North Strange Island
Middle Strange Island
Strange Island
BLACKJACK ISLAND
Cypress Creek
Sapp Prairie
PINHOOK SWAMP

Burnt Island
Cane Creek
Craven's Hammock
Bay Creek
Harper's Hammock
Tiddler's Island
Clinch Co.
Ware Co.
Rowell's Island

Tatum Creek
Jones Creek
Alligator Creek
FARGO   EDITH
RIVER
US 441

Clinch Co.
Echols Co.

GEORGIA
FLORIDA

SUWANNEE

Suwannoochee Creek

average water levels since construction of the Suwannee River Sill, a low earthen dam across the Suwannee River extending some 4.75 miles from the Pocket to Pine Island (76, 77).

The Okefenokee Swamp is the major source of both the Suwannee River and the north prong of the St. Marys River. Drainage into the Suwannee River accounts for approximately 85 percent of total runoff from the swamp and that of the St. Marys approximately 15 percent. The swamp runoff is not evenly distributed throughout the year and peak runoff is not always associated with peak rainfall. This is due largely to the water storing capacity of the peat in the swamp. Usually almost half the annual runoff occurs from January through April. This high spring runoff is apparently due to reduced evapotranspiration during cool winter and early spring months. Another period of peak runoff occurs in later summer and is associated with heavy rainfall during the summer months. There is usually a good correlation between average monthly stream-flow and mean monthly water level in the swamp (76, 77).

Climate

The climate of the Okefenokee Swamp is considered subtropical, intermediate between temperate and tropical. Throughout most of the year the region is influenced by tropical maritime air masses from the Gulf of Mexico and from the Atlantic Ocean. During winter, however, continental polar air masses may enter the area, resulting in severe cold in an otherwise mild winter.

In general, the seasons in the Okefenokee Swamp may be characterized as hot and wet in summer (May through September), warm and dry in fall (October and November), cool and moist in winter (December through February), and warm and moist in spring (March and April). Wet and dry are of course relative terms, and although fall is considered the dry season, 2 inches of rain is not uncommon. Significant temperature fluctuations generally occur only in winter. For example, during January temperatures may range from a low in the 20s to a high in the 80s.

Rainfall is not distributed evenly throughout the year; almost half the yearly average rainfall normally occurs during the summer from June through September. The dry season lasts from October through December. Rainfall during this period usually accounts for only 15 percent of mean annual total. Water level in the swamp is particularly susceptible to lowering when summer rainfall is below normal. This is exacerbated by normally low rainfall during the fall. Prolonged periods of significantly reduced rainfall may periodically cause severe droughts. Historically, 5 major droughts have been recorded in 1860,

12

1884, 1908, 1932, and 1954. Recently, the occurrence of major fires in 1909, 1933, and 1955 has been correlated with these periods of drought. The swamp typically recovers from low water levels when summer rainfall returns to normal (77).

Surface Geology

The coastal plain of southeastern Georgia, including the upland watershed of the Okefenokee Swamp, is characterized by intensively leached sandy soils that have a marked increase in the amount of clay with increasing depth. These soils are generally acidic and poor in nutrients. They were formed from marine sediments deposited during the Pleistocene (2 million years old) and Holocene (10 thousand years old) epochs.

Subsurface Geology

The subsurface geology of the swamp and surrounding uplands consists of an alternating series of permeable and impermeable marine sediments. These features are of significance in understanding the hydrology and origin of the swamp. The sediments underlying the swamp are believed to be continuous with those of the surrounding uplands. These include Pleistocene sands and underlying Pliocene (6 million years old) sands and gravels. These are of variable thickness, ranging from around 30 feet beneath the surface at Stephen C. Foster State Park west of the center of the swamp to 110 feet beneath the surface on the eastern border of the swamp. The Hawthorne Formation, which is several hundred feet thick and composed of sands, clays, and limestone of Miocene (22 million years old) origin, underlies the Pliocene sediments. The Hawthorne Formation is, in turn, underlain by the Ocala Limestone at a depth of about 500 feet. This limestone forms, in part, the principle artesian aquifer in the region. The Miocene and Pliocene deposits are relatively impermeable. The surface water table is confined above these (65).

Water Chemistry

The Okefenokee Swamp watershed is a naturally acidic ecosystem with an average pH value of 3.7. The water is stained a brownish tea color and is referred to as "blackwater." The dark color and the low pH values found there are due to the presence of large quantities of humic and fulvic organic acids (6), which are often called tannins or tannic acids. These acids result from the decomposition of organic materials, primarily plants.

Regions with unstained water or "whitewater" have surface sediments containing large amounts of clay, which can retain or trap the organic acids. The surface sediments of the Okefenokee Swamp and the surrounding coastal plain are chiefly sandy and cannot trap or retain the organic acids, which percolate into the groundwater and streams resulting in dark-stained water.

Although the swamp and the Suwannee and St. Marys rivers, when they exit from the swamp, are very acidic, the surrounding coastal plain rivers and streams are not. This is due (again) to changes in surficial geology. Downstream from the swamp the Suwannee and the St. Marys rivers flow over exposed calcareous deposits of Miocene and upper Oligocene origin (24). These deposits contribute to changes observed in water chemistry, such as higher concentrations of calcium and magnesium and more neutral pH values. The pH gradient found in the St. Marys River, for example, varies from 3.7 at the swamp exit to 4.8 some 35 kilometers downstream at the beginning of the Miocene limestone exposure, to an average pH 5.9 at 100 kilometers downstream.

Origin of the Swamp

The basin in which the Okefenokee Swamp rests is located on a marine terrace that formed from the transgression (high water levels) of Pleistocene seas probably during the Yarmouth interglacial period. Sea level was lowered during the following Illinoisian glacial period, exposing the sandy marine sediments. Trail Ridge, which forms the eastern boundary of the swamp, probably represents a large marine depositional sand bar or barrier island. The basin containing the swamp is considerably older than the swamp itself.

Considerable debate has transpired regarding the origin of the swamp. The traditional view held that the swamp originated as a saltwater lagoon that formed from the regression (low water levels) of Pleistocene seas. This saltwater lagoon was supposedly freshened by rainwater and subsequently invaded by vegetation, followed by the accumulation of peat. This view is not, however, supported by presently available evidence. Instead, the swamp appears to have originated more recently and under freshwater conditions. While the terrace upon which the swamp is located dates from the Pleistocene, radiocarbon dating of the oldest and deepest peat deposits indicates the swamp itself did not develop until fairly recent times, between six and seven thousand years ago.

Unlike the surrounding sandy uplands, peat is the predominant soil feature

14

in the swamp. Peat is plant matter that has been decomposed by microbial organisms. Because deposition occurs more rapidly than decomposition, peat accumulates and then absorbs water, causing the local water table to rise and the swamp to expand laterally. Essentially the swamp continues to hold water because it is underlaid by impermeable sediments through which water does not percolate to any significant extent.

Because of the large time span between the exposure of the marine terrace during the Pleistocene and the apparent recent development of peat, it is believed that the original terrace which contains the swamp was reworked by streams, winds, or other processes for a considerable time before the onset of peat accumulation and swamp formation. The oldest peats are found in the deepest basins in the swamp, suggesting that the primordial swamp was perhaps a series of disconnected or partially interconnected depressions with freshwater marshy vegetation (65).

Vegetation

The Okefenokee Swamp is far from being a monotonous marshland. On the contrary, compared to the surrounding uplands, the swamp exhibits a wide diversity of vegetational habitat types: *prairies, shrub swamps, blackgum forests, bay forests,* and *cypress forests* include the flooded or semiflooded portions of the swamp, while two additional habitat types, *islands* and the *uplands,* are dry.

Two prairie habitat types are recognized. Together they comprise approximately 21 percent of the swamp. Grass-sedge prairies are dominated by sedges, *Carex* sp.; panic grasses, *Panicum* sp.; beakrush, *Rhynchospora* sp.; broomsedge, *Andropogon virginicus;* giant chain fern, *Woodwardia virginica;* and *Sphagnum* sp. moss. Aquatic macrophyte prairies are characterized by emergent floating-leafed and submerged hydrophytes (water plants). This habitat is dominated by white water-lily, *Nymphea odorata;* ladies hatpin, *Eriocaulon compressum;* yellow water-lily, *Nuphar luteum;* never wet, *Orontium aquaticum;* floating heart, *Nymphoides aquaticum;* yellow-eyed grass, *Xyris smalliona;* pickerel weed, *Pontederia cordata;* redroot, *Lachnanthes caroliniana;* and bladderwort, *Utricularia* sp.

Shrub swamps make up approximately 34 percent of the swamp. They are dominated by hurrah bush, *Lyonia lucida;* fetterbush, *Leucothoe racemosa;* titi, *Cyrilla racemiflora;* sweet spire, *Itea virginica;* pepper bush, *Clethra alnifolia;* and dahoon, *Ilex cassine.*

Blackgum forests, located primarily in the northwestern portion of the swamp, constitute less than 6 percent of the swamp. Blackgum, *Nyssa sylvatica*

15

var. *biflora,* and to a lesser extent dahoon and pond cypress, *Taxodium ascendeus,* dominate the canopy. Red maple, *Acer rubrum,* and dahoon are the predominant understory plants.

Bay forests, which also make up less than 6 percent of the swamp, are dominated by loblolly bay, *Gordonia lasianthus;* red bay, *Persea borbonia;* and sweet bay, *Magnolia virginia.* Occasional pond cypress, blackgum, and slash pine, *Pinus elliottii,* also occur.

Cypress forests constitute approximately 23 percent of the swamp. Pure cypress forests are very limited in extent and consist almost entirely of cypress canopy with a sparse subcanopy or understory. Mixed cypress forests are more common. These are dominated by pond cypress, but loblolly bay, blackgum, and dahoon occur in the subcanopy.

Approximately 70 islands of varying size are located throughout the swamp and account for roughly 12 percent of the area. The substrate is typically sandy. The vegetation consists primarily of loblolly pine, *Pinus taeda;* slash pine; longleaf pine, *Pinus palustris;* and water oak, *Quercus niger.*

The typical uplands surrounding the swamp, including most of the watershed, are intensively managed slash pine forest plantations with an understory dominated by saw palmetto, *Serenoa repens;* small gallberry, *Ilex glabra;* and various forbs and grasses (48, 53).

Comment

Readers wishing further information on the Okefenokee Swamp and its natural history are referred to *The Okefenokee Swamp: Its Natural History, Geology, and Geochemistry,* edited by Cohen et al. (18). This book is a compendium of research papers dealing with the history and archeology, ecology, biochemistry, paleoecology, and geology and geomorphology of the swamp. The references included therein are extensive and provide access to nearly all the literature presently available on the Okefenokee Swamp.

History of Ichthyology in the Okefenokee Swamp

Scientific study of the fishes of the Okefenokee Swamp has not had a long history. Most of the focus of ichthyological study in the Southeast throughout the nineteenth and very early twentieth centuries was in Florida and the Carolinas. Scant attention was given to Georgia, and none, perhaps understandably, to the forbidding and inaccessible "land of the trembling earth"—the Okefenokee Swamp.

The only nineteenth-century ichthyologist to enter the Okefenokee Swamp was Charles Henry Bollman (64). In June 1889 he and a colleague, Bert Fesler, explored several of the lowland streams of southern Georgia. During this expedition Bollman was stricken with a fever from which he subsequently died. David Starr Jordan, a preeminent ichthyologist of the time, reports this as having occurred "in the Okefenokee Swamps in Georgia." Bollman's activities, however, appear to have been centered around Millen, Savannah, and Waycross. There is no evidence that he actually entered the boundaries of the swamp as we know it. In 1888 Charles H. Gilbert discussed specimens of the swampfish *Chologaster cornuta* from the Okefenokee Swamp "bordering the Ogeechee River near Millen." This is probably the Carolina Bay known as "Big Duke's Pond" or "Little Okefenokee Swamp." Perhaps it is also the "Okefenokee Swamp" in which Bollman contracted his fatal illness.

Other ichthyological studies were made quite near the Okefenokee Swamp before the turn of the century. W. J. Taylor collected to the west along the Alapaha River (64); to the south, A. J. Woolman made collections along the Sante Fe River (64). Both of these rivers are tributaries of the Suwannee.

It was not until 1920 that the first published records of the fishes of the Okefenokee Swamp became available. Ephraim Laurence Palmer and Albert Hazen Wright were among the biologists from Cornell University who extensively studied the flora and fauna of the swamp from 1912 throughout the 1920s. Their 1920 publication, *A Biological Reconnaissance of the Okefenokee Swamp in Georgia: The Fishes* (64), remained the definitive account of the fishes of the Okefenokee Swamp for the next 60 years. Both Palmer and Wright were naturalists with broad interests. Palmer was well known as a botanist and limnologist; Wright is perhaps best known for his in-depth studies of amphibians. Also associated with the Cornell expeditions was the young naturalist Francis Harper, who was involved in studies of the Okefenokee Swamp and its people over the next 40 years. Although best known as a mammalogist, Harper made a number of collections of fishes for Cornell University.

In retrospect, the Cornell expeditions and Palmer and Wright's publication may have preempted significant subsequent studies of the Okefenokee Swamp. During the 50 years following their work, no substantive scientific investigations were undertaken, and only a few biologists made significant collections in the region. We record here those whose collections can be substantiated from museum records: R. T. Berryhill in 1924, T. Reichelderfer in 1935, M. S.

Verner, Jr., in 1936, B. Cadbury in 1937, C. B. Obrecht and M. Godfrey in 1941, H. A. Carter in 1941–42, T. Rodenberry in 1941, Southern Piedmont and Coastal Plain Aquatic Survey in 1941, R. J. Fleetwood in 1947, E. Cypert in 1960–63, T. Cavender in 1965, and J. Bohlke in 1966. From 1970 to 1973 Daniel R. Holder of the Georgia Department of Natural Resources intensively studied the upper portion of the Suwannee River and portions of the Okefenokee Swamp. In 1976 Robert D. Gassaway made detailed studies of several tributary streams of the Suwannee River in the Okefenokee watershed.

Since 1978 the University of Georgia Museum of Natural History has undertaken major surveys of the vertebrate fauna of the Okefenokee Swamp (48). This work, in conjunction with broad ranging ecosystem analysis conducted by scientists at the University of Georgia and other institutions, continues today (18).

Comparison with Regional Fauna

The ichthyofauna of the Okefenokee Swamp is similar to that of adjacent southeastern river drainages, although it is less diverse (48). The fauna comprises 36 fishes representing 14 families; hence most families in the Okefenokee Swamp are only represented by one or two species. Exceptions are the sunfishes, catfishes, and topminnows.

Outdoor people familiar with south Georgia and north Florida waters will recognize the largemouth bass, warmouth, flier, chain pickerel, gar, and mudfish. The plucky bluegill, common in ponds, lakes, and streams elsewhere, is rarely encountered in the Okefenokee Swamp. The large expansive marshes are populated by small, often-overlooked species such as the starhead topminnow, mosquitofish, and pygmy sunfish. Although some may be unfamiliar, all Okefenokee Swamp fishes also live in surrounding coastal plain waters.

Remarkable features of the fish assemblage of the Okefenokee Swamp are the complete absence of minnows (family Cyprinidae) and the absence of various species of sunfishes (families Elassomidae and Centrarchidae). Minnows are an important group of freshwater fishes; at least 69 species inhabit streams and lakes in Georgia alone (23). Nine species may be collected from the river drainages surrounding the Okefenokee Swamp and might be expected to occur in the abundant habitats of either the swamp or streams draining from it. The golden shiner (*Notemigonus crysoleucas*), ironcolor shiner (*Notropis chalybaeus*), pugnose minnow (*N. emiliae*), sailfin shiner (*N. hypselopterus*), Ohoopee shiner (*N. leedsi*), taillight shiner (*N. maculatus*), coastal shiner (*N. petersoni*), weed shiner (*N. texanus*), and blacktail shiner (*N. venustus*) all inhabit waters surrounding the Okefenokee Swamp. Sunfishes found around, but not in, the Okefenokee Swamp include the banded pygmy sunfish (*Elassoma zonatum*), redbreasted sunfish (*Lepomis auritus*), red-ear sunfish (*L. microlophus*), Suwannee bass (*Micropterus notius*), and white crappie (*Pomoxis annularis*).

One possible explanation for the limited variety of fauna in the Okefenokee Swamp, as compared with other blackwater coastal plain waters, is the highly acidic nature of the swamp. The average pH of the Okefenokee Swamp is 3.7, with values as low as 3.1 having been recorded (9). Distributional records for the surrounding river drainages indicate that cyprinids are not found in areas with an average pH value of less than 4.5. Our observations indicate that adult coastal shiners, *Notropis petersoni*, actively avoid acidic water which they would encounter when moving upstream in the Suwannee and St. Marys rivers toward the swamp. Coastal shiners cannot survive when transferred to low pH water (approximately 3.7) from the swamp interior or from either the Suwannee or St. Marys rivers. The argument for pH or pH-associated limitation can best be summarized by the following points: no natural barriers exist to prevent migration upstream from downstream populations; adult cyprinids can detect and avoid acidic waters, and they fail to survive at pH values less than 4.2; and populations of cyprinids have not been found in habitats with average pH values less than 4.5.

Although the fauna of the Okefenokee may be restricted in kind, the waters are not necessarily unproductive. Combined production of small marsh fishes may, in some years, exceed levels observed for other North American lake and stream fish assemblages. Fishing is a popular activity in the Okefenokee Swamp, and many an angler has been well rewarded by a day's trip to its quiet, sometimes mysterious, but always beautiful waters.

How to Use the Keys

To the uninitiated, a key might first appear rather confusing. It's not. Keys are the standard device of biologists for identifying species. Keys are usually fairly straightforward, and with a little practice prove to be extremely helpful. Two keys are included in this book. The first is a key to the families of fishes and will assist the reader in identifying the family to which a particular fish belongs. Within each family a second key will help identify the individual species.

A key is composed essentially of a series of alternative statements, either *a* or *b*, referring to particular morphological features. By starting at the beginning and selecting the obvious alternative, the reader is eventually directed by the key to the appropriate family and, similarly, within the family to the appropriate species.

For example, imagine you have caught a fish that you will learn is a chain pickerel. How do you find out? By using the keys. For the sake of this example use the illustration of the chain pickerel on page 42. Starting with the Key to the Families on page 21, statement 1 provides two alternatives, a or b, relating to tail shape. The tail is clearly homocercal. The key says to go to statement 3. Here two alternatives relating to body shape and presence or absence of pelvic fins are given. The obvious choice is 3b which says "go to 4." Statements 4a and b provide two other alternatives of which 4b is correct because of the absence of an adipose fin and barbels. This directs you to 5. In 5 you are asked to choose between (a) a single dorsal fin with soft rays and (b) two dorsal fins or, if one, the one having spinous rays. Here again the choice is obvious—5a, which directs you to statement 6. Looking at the head of the fish, you see it has scales. You select 6b, and this statement directs you to 7. Statement 7 provides two alternatives. Is the tail forked? Yes. Are the jaws elongate, forming a ducklike snout? Yes. The fish you have caught is in the family Esocidae. Now, turn to the family Esocidae on page 24. Here we have a short key to distinguish between the two species of pickerel found in the Okefenokee Swamp. Comparing the two alternatives— particularly the angle of the dusky bar beneath the eye, the long snout, and chainlike markings on the side—it is clear that you have just "keyed out" or identified *Esox niger*, the chain pickerel.

20

Key to the Families

1a. Tail heterocercal or abbreviate heterocercal.   Go to 2

 b. Tail homocercal.   Go to 3

2a. Snout greatly elongate; ganoid body scales rhomboid and nonoverlapping, dorsal fin base short; no gular plate.   Lepisosteidae, p. 23

 b. Snout not elongate; scales cycloid and overlapping, dorsal fin base long with more than 45 rays, extending over half the length of the body.   Amiidae, p. 23

3a. Pelvic fins absent; caudal fin confluent with dorsal and anal fins over posterior part of body; body shape "eel-like."   Anguillidae, p. 23

 b. Pelvic fins present; body not "eel-like."   Go to 4

4a. Adipose fin present; barbels present around mouth; body without scales.   Ictaluridae, p. 25

 b. Adipose fin absent; barbels absent; body with scales.   Go to 5

5a. Single dorsal fin composed entirely of soft segmented rays.   Go to 6

 b. Either two separate dorsal fins or, if one, the anterior unsegmented rays are spinous.   Go to 10

6a. Head (including cheek) without scales.   Catostomidae, p. 25

 b. Head with at least some scales, the cheek always with scales.   Go to 7

7a. Caudal fin forked; jaws elongate, forming a ducklike snout.   Esocidae, p. 24

 b. Caudal fin rounded; jaws not elongate and not forming a ducklike snout.   Go to 8

8a. Premaxillary bones protractile; lower jaw projecting, frenum absent.

Go to 9

b. Premaxillary bones not protractile; lower jaw not projecting, frenum present.

Umbridae, p. 24

9a. Third anal ray unbranched; anal fin of males modified into an intromittent organ, the gonopodium.

Poeciliidae, p. 27

b. Third anal ray branched, anal fin of males not modified.

Cyprinodontidae, p. 26

10a. Anus located on throat region anterior to pectoral fins.

Aphredoderidae, p. 26

b. Anus located on thoracic or abdominal region posterior to pectoral fins.

Go to 11

11a. Two separate dorsal fins.

Go to 12

b. Dorsal fin single or incompletely separated.

Go to 13

12a. Base of anal fin longer than base of soft dorsal fin.

Atherinidae, p. 27

b. Base of anal fin shorter than base of soft dorsal fin.

Percidae, p. 30

13a. Dorsal fin spines between 3 and 5 in number.

Elassomidae, p. 28

b. Dorsal fin spines more than 6 in number.

Centrarchidae, p. 28

Fish Families
and Keys to the Species

Lepisosteidae · Gars

Gars are long, slender, cylindrical fish with an elongate snout containing sharp teeth. The body is covered with heavy, bony (ganoid) scales. Gars are a very primitive group of fishes most closely related to fishes known only from fossils and almost all of which have been extinct for over 200 million years. Gars are generally restricted to the temperate and tropical regions of eastern North America. They inhabit warm, slow-moving waters of lowland rivers and lakes. Tolerant of saline conditions, they also occur in southern estuarine waters. There is but a single genus, *Lepisosteus*, and 7 species, five of which occur in the southeastern United States. Only *Lepisosteus platyrhincus* occurs in the Okefenokee Swamp.

Amiidae · Bowfins

Bowfins are moderately large, stout-bodied, predatory fishes. They are the only living representatives of the order AMIIFORMES, a primitive group of fishes known largely from fossils. These fishes were common throughout the Triassic, Jurassic, and Cretaceous periods some 230 to 80 million years ago. The family is represented by a single living species, *Amia calva*, which occurs throughout lowland fresh waters of much of eastern North America, including the Okefenokee Swamp.

Anguillidae · Freshwater Eels

Eels have extremely elongate bodies, almost round in cross section. Approximately 32 families of true eels are found worldwide; however, all but the Anguillidae are marine. Freshwater eels are catadromous, that is, they breed at sea. They enter fresh waters early in their development and remain there throughout their lives until they return to the sea to spawn and die. Although there are 16 species in the genus *Anguilla*, occurring throughout much of the world, only a single species, *Anguilla rostrata*, is found in North America.

Esocidae · Pikes

This family includes the pike, muskellunge, and pickerels. These are medium to moderately large fishes with cylindrical bodies, a duckbill-shaped snout, and forked tail. The family contains a single genus, *Esox*, which is distributed throughout most of the northern hemisphere. They occur in lakes, ponds, rivers, streams, and to a lesser extent in brackish water. Five species are described worldwide, four of which occur in North America. Two species, both pickerel, live in the Okefenokee Swamp region.

Key to the Species of Pickerel

a. Dusky bar beneath eye angled downward and slightly backward; snout shorter, with distance from tip of snout to center of eye equal to or less than distance from center of eye to posterior edge of opercle; adult with irregular vertical bars on sides; 12–13 branchiostegal rays on each side.
Esox americanus

b. Dusky bar beneath eye vertical; snout longer, with distance from tip of snout to center of eye equal to or greater than distance from center of eye to posterior edge of opercle; adult with chainlike markings on side; 14–16 branchiostegal rays on each side. *Esox niger*

Umbridae · Mudminnows

Mudminnows are small fishes with rounded caudal fins. They occur in slow-moving waters with abundant vegetation in Europe, Asia, and subtropical to Arctic North America. There are 3 genera, including 5 species worldwide. Only one species, *Umbra pygmaea*, is found in the Okefenokee Swamp.

Catostomidae · Suckers

Suckers are small to moderately large fishes with subcylindrical to laterally compressed bodies. In North America the family is represented by 10 living genera. Suckers are bottom-dwelling fishes with a wide variety of habitat preferences. Two species, representing two genera, occur in the Okefenokee Swamp.

Key to the Species of Suckers

a. Snout extending well beyond the upper lip; posterior margin of dorsal fin straight or slightly concave; scales on side with dark spots on base forming parallel lines. *Minytrema melanops*

b. Snout extending only slightly beyond upper lip; posterior margin of dorsal fin convex; scales on side without dark spots on base. *Erimyzon sucetta*

Ictaluridae · Catfishes

These catfishes, including the bullheads and madtoms, occur exclusively in North America. They are small to large fishes with large, flattened heads. They have whiskerlike barbels, no scales, and a well-developed spine at the origin of both dorsal and pectoral fins. These fishes are characteristic of subtropical and temperate, slow-moving rivers, streams, and adjacent waters. There are 5 genera, including 39 species. Two genera and 5 species are found in the Okefenokee Swamp.

Key to the Species of Catfishes

1a. Adipose fin short, its posterior margin free and well separated from the caudal fin. *Ictalurus*, go to 2

b. Adipose fin fleshy, long, and keel-like either connected to caudal fin or separated from it by a shallow notch. *Noturus*, go to 4

2a. Caudal fin not deeply notched. Go to 3

b. Caudal fin deeply notched. *Ictalurus punctatus*

3a. Chin barbels whitish; rear margin of caudal fin nearly straight; 24–27 anal fin rays. *Ictalurus natalis*

  b. Chin barbels black or gray; rear margin of caudal fin slightly notched; 22–23 anal fin rays. *Ictalurus nebulosus*

4a. Body and fins not freckled with dark spots; pectoral spine long and stout, half as long as head length, without serrations on anterior edge; body short, its standard length less than 4.5 times its maximum depth. *Noturus gyrinus*

  b. Body and fins freckled with dark spots; pectoral spines short and slender, less than a third of head length, with slight serrations on anterior edge; body elongate, its standard length more than 4.5 times its maximum depth. *Noturus leptacanthus*

Aphredoderidae · Pirate Perches

Pirate perches are small perchlike fishes. The family Aphredoderidae has only one genus, which has a single species, *Aphredoderus sayanus*. This species is confined to the eastern United States. The closest relatives of the pirate perches are the trout perches and cave fishes; however, neither of these groups occurs in the Okefenokee Swamp region.

Cyprinodontidae · Killifishes

Killifishes are small, stout-bodied fish. The head and mouth are usually adapted for surface feeding. Most killifishes prefer shallow fresh water, but many are known to enter brackish water and even seawater. They occur widely in temperate and tropical waters worldwide and are abundant in the south-eastern United States. Forty-five genera and approximately 300 species exist worldwide. Two genera and 4 species are found in the Okefenokee Swamp.

Key to the Species of Killifishes

1a. Teeth in more than 1 row; teeth in outer row may be large and in inner row small; no well-defined dark spot at base of caudal fin.
Fundulus, go to 2

b. Teeth in 1 row; well-defined dark spot at base of caudal fin.
Leptolucania ommata

2a. Well-defined dark teardrop below eye.   *Fundulus lineolatus*

  b. No well-defined dark teardrop below eye.   Go to 3

3a. Narrow, middorsal gold line anterior to dorsal fin; elongate gold spot above and behind eye.   *Fundulus cingulatus*

  b. No narrow, middorsal gold line; no elongate gold spot above and behind eye.   *Fundulus chrysotus*

Poeciliidae · Livebearers

Livebearers are so named because, unlike most other fishes who shed eggs that develop outside the female, the young livebearers develop inside the female and are born live. Livebearers are among the smallest fishes in North America. They are strictly New World in their distribution, ranging from central United States south into South America. Six genera exist, including 20 species in North America. Two genera, each with a single species, occur in the Okefenokee Swamp.

Key to the Species of Livebearers

  a. Dark longitudinal band and vertical bars along the sides; dark blotches on dorsal and caudal fins of males and on caudal and anal fins of females.
Heterandria formosa

  b. No dark longitudinal band or vertical bars along the sides; no dark blotches on fins.   *Gambusia affinis*

Atherinidac · Silversides

Silversides are small, thin, elongate, and almost transparent silvery fishes with a pronounced lateral stripe. The family includes 50 genera and approximately 170 named species worldwide. While the majority of the species occur in temperate to tropical marine, coastal, and brackish waters, a few live in fresh water. Three species are found in fresh water in the southeastern United States, but only a single species, *Labidesthes sicculus*, inhabits the Okefenokee Swamp.

Elassomidae · Pygmy Sunfishes

Superficially, pygmy sunfishes resemble sunfishes in the family Centrarchidae, and until recently they were regarded as members of that family. Major differences, however, are discernible in morphology and reproductive behavior between them, and most experts regard pygmy sunfishes as a separate family (see Key to the Families).

Pygmy sunfishes are small, diminutive "sunfishes," ranging from 1½ to 3 inches in length. They include at least three species, which are primarily restricted to the southeastern Atlantic and Gulf coast drainages. Two species occur in the Okefenokee region.

Key to the Species of Pygmy Sunfishes

  a. Top of head with embedded scales; 8–10 dorsal rays; 4–5 anal rays.
Elassoma evergladei

  b. Top of head without scales; 10–13 dorsal rays; 6–8 anal rays.
Elassoma okefenokee

Centrarchidae · Sunfishes

Sunfishes are small- to moderate-sized, generally deep-bodied, and laterally compressed fishes. The dorsal and anal fins consist of spinous and soft rays. These fishes include the sunfishes (or bream to most southerners), crappie, and bass. They include some of the most highly colored and attractive fishes as well as the most popular game fishes in North America. The 10 genera and 30 species that constitute the family are limited in their distribution to North America, where they inhabit lakes, ponds, rivers, and slow-moving streams. Six genera and 11 species inhabit the Okefenokee Swamp.

28

Key to the Species of Sunfishes

1a. More than 3 anal spines. Go to 2

 b. Three anal spines. Go to 4

2a. Less than 10 dorsal spines. *Pomoxis nigromaculatus*

 b. Ten or more dorsal spines. Go to 3

3a. Tail rounded; 5 anal spines. *Acantharchus pomotis*

 b. Tail truncate or slightly emarginate; 7 or 8 anal spines.
Centrarchus macropterus

4a. Tail rounded. *Enneacanthus*, go to 5

 b. Tail emarginate or forked. Go to 7

5a. Body with 6 prominent, vertical dark bands, the first passing through the eye, the third onto the anterior edge of the dorsal fin; leading edge of pelvic fins light-colored, usually orange-red in life.
Enneacanthus chaetodon

 b. Vertical dark bands variable, none passing onto anterior edge of dorsal fin; leading edge of pelvic fins not noticeably light-colored. Go to 6

6a. Opercular spot round, size greater than half the diameter of the eye; 5–8 distinct, vertical dark bars on the side; no blue spots on sides, head, and fins. *Enneacanthus obesus*

 b. Opercular spot angular, size less than half the diameter of the eye; vertical bars narrow or indistinct; blue spots on sides, head, and fins.
Enneacanthus gloriosus

7a. Spinous and soft portions of dorsal fin well connected with no deep notch between; body compressed and deep (sunfishes of the genus *Lepomis*).
Go to 8

 b. Spinous and soft portions of dorsal fin divided by a deep notch; body elongate and not much compressed. *Micropterus salmoides*

29

8a. Three dark bars radiating backward from the eye; mouth large, extending beyond pupil of the eye; teeth present on tongue.   *Lepomis gulosus*

  b. No dark bars behind the eye; mouth smaller, not extending beyond pupil of eye; no teeth on tongue.   Go to 9

9a. Opercular bone stiff to margin; longitudinal rows of light spots along sides.   *Lepomis punctatus*

  b. Opercular bone flexible at posterior margin; no rows of light spots along sides.   Go to 10

10a. Vertical bars (3–6) on sides; black spot on posterior soft dorsal fin; no greenish white margin around opercular flap.   *Lepomis macrochirus*

  b. No vertical bars; no dark spot on soft dorsal fin; greenish white margin on opercular flap.   *Lepomis marginatus*

Percidae · Perches and Darters

Perches and darters are elongate, small- to moderate-sized fishes. They have two well-separated dorsal fins with the front fin made up of spiny rays. The family includes two subfamilies, the perches (subfamily Percinae) and the darters (subfamily Etheostomatinae). Perches occur in North America, Europe, and northern Asia. Darters are restricted to North America. The family consists of 9 genera and approximately 120 species. Two species, both darters, live in the Okefenokee Swamp.

Key to the Species of Darters

  a. Belly unscaled except for a row of enlarged or modified scales on the midventral line; 11–15 vertically elongate, diamond-shaped blotches along the sides; dorsum has 6–8 squarish blotches; lateral line straight and complete to base of tail.   *Percina nigrofasciata*

  b. Belly scaled, no enlarged midventral scales; irregular lateral band consisting of scattered, irregular blotches present but no diamond-shaped blotches; no defined blotches on dorsum; lateral line arched anteriorly and incomplete, not extending to base of tail.   *Etheostoma fusiforme*

List of the Species

ORDER SEMIONOTIFORMES
  Family Lepisosteidae
    Lepisosteus platyrhincus   Florida gar

ORDER AMIIFORMES
  Family Amiidae
    Amia calva   bowfin

ORDER ANGUILLIFORMES
  Family Anguillidae
    Anguilla rostrata   American eel

ORDER SALMONIFORMES
  Family Esocidae
    Esox americanus   redfin pickerel
    Esox niger   chain pickerel
  Family Umbridae
    Umbra pygmaea   eastern mudminnow

ORDER CYPRINIFORMES
  Family Catostomidae
    Erimyzon sucetta   lake chubsucker
    Minytrema melanops   spotted sucker

ORDER SILURIFORMES
  Family Ictaluridae
    Ictalurus natalis   yellow bullhead
    Ictalurus nebulosus   brown bullhead
    Ictalurus punctatus   channel catfish
    Noturus gyrinus   tadpole madtom
    Noturus leptacanthus   speckled madtom

ORDER PERCOPSIFORMES
  Family Aphredoderidae
    Aphredoderus sayanus   pirate perch

ORDER ATHERINIFORMES
  Family Cyprinodontidae
    Fundulus chrysotus   golden topminnow
    Fundulus cingulatus   banded topminnow
    Fundulus lineolatus   lined topminnow
    Leptolucania ommata   pygmy killifish
  Family Poeciliidae
    Gambusia affinis   mosquitofish
    Heterandria formosa   least killifish
  Family Atherinidae
    Labidesthes sicculus   brook silverside

ORDER PERCIFORMES
  Family Elassomidae
    Elassoma evergladei   Everglades pygmy sunfish
    Elassoma okefenokee   Okefenokee pygmy sunfish
  Family Centrarchidae
    Acantharchus pomotis   mud sunfish
    Centrarchus macropterus   flier
    Enneacanthus chaetodon   blackbanded sunfish
    Enneacanthus gloriosus   bluespotted sunfish
    Enneacanthus obesus   banded sunfish
    Lepomis gulosus   warmouth
    Lepomis macrochirus   bluegill
    Lepomis marginatus   dollar sunfish
    Lepomis punctatus   spotted sunfish
    Micropterus salmoides   largemouth bass
    Pomoxis nigromaculatus   black crappie
  Family Percidae
    Etheostoma fusiforme   swamp darter
    Percina nigrofasciata   blackbanded darter

Fishes of the
Okefenokee Swamp

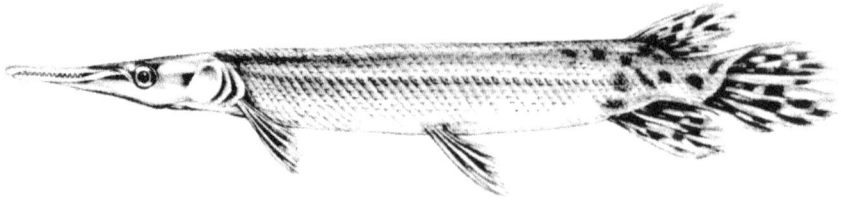

Florida Gar

Lepisosteus platyrhincus DeKay

Other Local Names.   Billfish, billy gar, gar, shortnose gar, spotted gar.

Scientific Name.   *Lepisosteus*, from the Greek meaning "bony scale"; *platyrhincus*, also from the Greek referring to the flattened snout.

Description.   The Florida gar is a large fish, ranging in length from 12 to 36 inches—one caught at Mixon's Ferry, however, was reported to be over 4 feet long (64). The Florida gar is a long, slender fish with a cylindrical body and an elongate snout containing numerous sharp teeth. The head and snout lack scales but are covered by bony plates. A bony isthmus is present. The body is covered by heavy, nonoverlapping bony (ganoid) scales that form a nearly impenetrable covering. The scales are rhomboid in shape and are arranged in oblique rows. The caudal fin is rounded. The tail itself is abbreviate heterocercal with the axis of the vertebral column extending dorsally into the tail. The dorsal and anal fins are posterior in position and nearly opposite one another. Females are typically larger than males and have a proportionally longer snout.

The back and sides of the Florida gar are dark olive brown. The belly is lighter and may be patterned with dark stripes. Most specimens have a variable pattern of dark spots, particularly on the head, the anterior part of the body, and the fins. Bright orange specimens (68) have been reported in Florida, but this color phase has not been found in the Okefenokee Swamp.

Similar Species.   None. This is the only gar in the Okefenokee Swamp, yet both this species and the longnose gar (*L. osseus*) occur throughout the rest of both the St. Marys and Suwannee rivers. The Florida gar may be distinguished from the longnose gar by the much longer snout of the latter.

Distribution.   The Florida gar is found throughout Florida east of the Ochlockonee River and across the extreme southern portion of Georgia north to the Savannah River. It occurs in both the St. Marys and Suwannee rivers and also in much of the Okefenokee Swamp.

Habitat.   This species is an inhabitant of lakes and the sloughs and oxbows of rivers, streams, and estuaries. It is invariably associated with submerged vegetation and is infrequently encountered in coastal marine waters. In the Okefenokee Swamp the Florida gar lives in a variety of habitats with heavy vegetation, including canals, lakes, prairies, boat runs, creeks, and sloughs.

The Florida gar is not a significant component of most local fish faunas (62). In the Okefenokee Swamp it totals less than 1 percent of collections (33, 37). Gars are gregarious and swim in groups of 2 to 10.

Biology.   The Florida gar feeds mostly on smaller nongame and game fishes, primarily bass and bream. Various crustaceans and aquatic insects make up the remainder of the diet (40, 80). The species feeds all day but is most active at dusk and dawn.

The fish typically spawns in March and April, but spawning dates range from January to as late as October (62). More than 10,000 eggs may be shed. Little information regarding breeding behavior and development is available for this species.

Comments.   Because a hook will rarely penetrate the hard, bony jaw, the Florida gar is not frequently taken by sportsmen. The gar has little commercial value, although the roe (eggs) are sometimes used for bream bait. The roe is toxic to mammals—including humans—and *should not be eaten.*

In appropriate habitat the Florida gar may be observed basking beneath the surface. A gar will frequently "break water," rising to the surface, opening its mouth, and snapping it shut again in order to "gulp air." This behavior allows the fish to fill its swim bladder with air in order to augment gill respiration. The swim bladder is physostomous (connected to the pharynx by a duct). Most other fishes also have a swim bladder but have no duct connecting it with the pharynx (physoclistous). In all fishes that have one, the swim bladder functions as a density control (or flotation) device.

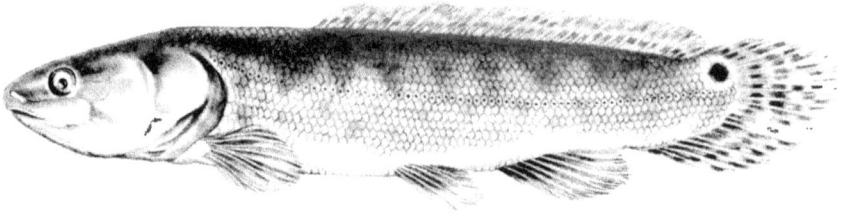

Bowfin

Amia calva Linnaeus

Other Local Names.   Cypress trout, dogfish, grindle, mudfish, scaly cat, spottail.

Scientific Name.   *Amia,* from the Greek for "tunny," a fish; *calva,* from the Latin meaning "bold" or "smooth," in reference to its unscaled head.

Description.   The bowfin is a moderate-sized fish, with adults ranging up to 24 inches in length. The largest known specimen is almost 36 inches. The bowfin is stout-bodied and nearly cylindrical in shape. The large, unscaled head is rounded,bearing tubular anterior nostrils. The mouth is large with many sharp, pointed teeth. A bony gular plate is present between the lower jaws. The body is covered with large, overlapping cycloid scales. The dorsal fin is spineless and long, extending more than half the length of the body. The caudal fin is rounded. Like the gar's, the tail is abbreviate heterocercal.

The coloration on the back is mottled dark olive and is lighter on the sides and belly. The dorsal and caudal fins are dark olive with darker bands or bars. The anal and paired fins are bright green. At the base of the upper caudal fin rays, there is a black, oval spot with a yellow-to-orange surrounding halo. This marking is most pronounced in juveniles and males and may be inconspicuous or absent in females. The young may be brightly colored with orange and blue on the head and fins.

Similar Species.   The bowfin is very distinctive, but juvenile fish might be confused with mudminnows, which have a much shorter dorsal fin.

Distribution. The bowfin occurs throughout much of the eastern United States—except the Appalachian Highlands—from Quebec west to the Mississippi River drainage and south to the Gulf of Mexico. It is distributed widely throughout the coastal plain of Georgia and Florida, including both the St. Marys and Suwannee rivers and the Okefenokee Swamp watershed.

Habitat. The bowfin lives in sluggish, clear, lowland fresh waters with heavy plant growth, including the backwater pools of rivers, swamps, and lakes as well as sloughs, oxbows, ditches, and borrow pits. In the Okefenokee Swamp it has been taken in lakes, prairies, sloughs, and small streams and creeks.

In smaller creeks the bowfin constitutes a small part of fish populations. For example, at three sites along Alligator Creek and Jones-Tatum Creek, the bowfin made up between 1 and 6 percent of the total fishes taken in traps (33). The fish is very common, however, in the region of the Sill and the Suwannee River below the Sill. In one study (37), the bowfin accounted for 27 percent of the fish taken at the spillway at the East Prong of the Suwannee River and 30 percent of the fish falling over the spillway.

Biology. The adult bowfin is predaceous. Its food includes other smaller fishes, including game and panfish, amphibians, crayfish, and to a lesser extent invertebrates (8, 38, 49, 80). It feeds during the day but is probably most active at night.

In northern Florida the bowfin breeds from February through late spring (55). The male bowfin is territorial and constructs a large (1–2 foot) circular nest in vegetation. After a brief courtship, the female lays between 2,000 and 64,000 eggs (13). The male aggressively defends the nest and eggs. The eggs are adhesive and remain attached to the nest until hatching takes place in 8–10 days. Fry remain in the nest another 8–10 days. After leaving the nest, the fry, still protected by the male, remain in a compact swarm until they are approximately 4 inches long and about a month old.

Comments. The bowfin is frequently caught by fishermen. It readily takes bait and lures and must be fished on the bottom. Because of its strength and endurance, the bowfin provides considerable sport when taken on light tackle. The bowfin is generally considered a poor food fish, though the taste may be improved by smoking or baking the fish in a highly spiced marinade.

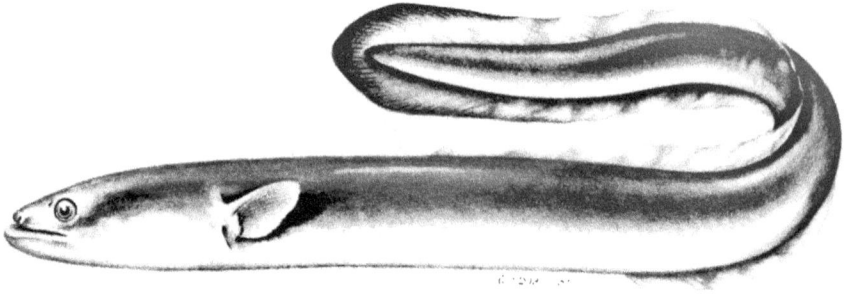

American Eel

Anguilla rostrata (Lesueur)

Other Local Names. Eel, fish eel, river eel, snake fish.

Scientific Name. *Anguilla,* from the Latin meaning "eel"; *rostrata,* also from the Latin meaning "beaked."

Description. The American eel is a very distinctive fish because of its exceedingly long body that is almost round in cross section. Most adult female eels range from 12 to 36 inches, with a maximum length of 48 inches. Males are smaller and rarely exceed 20 inches. The head is elongate and has a large mouth with numerous teeth and a protruding lower jaw. The scales are embedded and small—so small that the eel appears scaleless. There is a single, small gill opening in front of each pectoral fin. There are no pelvic fins. The dorsal, caudal, and anal fins are continuous. The tail is blunt and rounded.

The coloration of eels varies during their life cycle and is dependent upon habitat. By the time American eels begin to enter coastal rivers, they are approximately one year old and grayish green in color. During their years of sexual immaturity in fresh waters their color ranges from yellow to olive brown. The back is typically darker than the belly. When sexually mature and ready to return to the sea to breed, the back is dark metallic, the belly a light silver. Only the two latter color phases would be expected in eels in the Okefenokee Swamp.

Similar Species. The American eel is not easily confused with any other fish. It does, however, superficially resemble two large amphibians—the siren and the

38

amphiuma—that live in the Okefenokee Swamp. The siren has small front legs and external gills. The amphiuma has small front and hind legs and is often referred to locally as the snake eel.

Distribution.   The American eel occurs throughout most of eastern North America from the coast of Labrador south to South America including the West Indies. This fish is found in both the St. Marys and Suwannee rivers.

Habitat.   In fresh water, eels prefer deep pools of rivers, creeks, lakes, and ponds with muddy bottoms. During the day they hide beneath rocks, logs, or other submerged debris. They frequently prefer waters with vegetation. While eels are common in both the St. Marys and Suwannee rivers, they are uncommon in the Okefenokee Swamp, having been recorded only from Billy's Lake and Red Bluff Slough.

Biology.   Eels are voracious eaters and feed primarily at night. They consume a wide variety of animal food (both living and dead), including fishes, amphibians, and various invertebrates, principally crustaceans (36, 55).

In general, very little is known about the breeding biology of eels. Sexually mature individuals migrate to the sea to spawn after spending as many as 5–20 years in fresh waters. The breeding ground of the American eel is the Sargasso Sea or areas south of it (78, 87). Freshly hatched eels are transparent, almost leaf-shaped, and so unlike eels that they were once thought to be a different kind of fish, *Leptocephalus*. The leptocephalus larvae drift with ocean currents into American waters by the following winter. By this time they are approximately 3 inches long and ready to transform into "glass eels" or "elvers." At this stage the young resemble the adults but almost completely lack pigmentation. Eels penetrate the mouths of rivers as elvers. Despite their voracious appetite, eels grow slowly (they may reach only 6–12 inches in the first 5–8 years) and also mature slowly (5–20 years).

Comments.   Despite the American eel's reputation as a good fighter on light tackle, popular prejudice exists against this fish in the United States. This is unfortunate, for eels are an excellent food fish. The American eel supports a limited commercial fishery in the United States and Canada, primarily through export to Europe (25, 29).

Redfin Pickerel

Esox americanus Gmelin

Other Local Names.   Banded pickerel, grass pickerel, jack-fish.

Scientific Name.   *Esox,* from the Latin for "pike"; *americanus,* from the Latin meaning "from America."

Description.   The redfin pickerel is the smallest of the pike family; adults are usually 10–12 inches in length and rarely attain a maximum of 15 inches. The redfin pickerel has an elongate cylindrical body that is slightly compressed laterally. The head is elongate with cheeks and opercle more or less fully scaled. The snout is duckbilled and shorter than that of the chain pickerel; the distance from the tip of the snout to the center of the eye is equal to or less than the distance from the center of the eye to the posterior margin of the opercle. Branchiostegal rays along the lower margin of the gill cover numbers 12–13 on each side. The dorsal fin is located posterior, opposite the anal fin. The caudal fin is forked. Scales in the lateral line range in number from 92 to 117.

The back and sides are olive to yellowish brown. The sides are often marked with irregular, dark, wavy vertical bars. The ventral surface is milky white. A dusky bar extends downward and backward from the eye. The caudal, anal, and paired fins are frequently orange-red in color, hence the fish's common name.

Similar Species.   The redfin pickerel may be confused with the chain pickerel (*E. niger*), which also occurs in the Okefenokee Swamp. In *E. americanus* the dusky bar beneath the eye is angled slightly backward; in *E. niger* it is vertical. The snout in *E. americanus* is shorter than in *E. niger.* The adult *E. americanus* has irregular vertical bars on the side; *E. niger* has chainlike reticulations instead.

Distribution. This species ranges throughout the eastern United States, except in the Appalachian highlands and adjacent areas of extreme southern Canada. It occurs in the southern Great Lakes and south through the Mississippi Valley, east along the Gulf Coast, and up the Atlantic Coastal Plain. It is common in both the St. Marys and Suwannee rivers and in the Okefenokee Swamp.

Habitat. The redfin pickerel inhabits quiet acidic waters with heavy plant growth, including the quiet pools of rivers and streams, canals, boat runs, sloughs, and prairies. It occurs most commonly on or near the surface along the edge of bodies of water. Redfin pickerel make up less than 1 percent of the fish caught in studies at the East Prong Spillway of the Suwannee River (37).

Biology. Like other members of the pike family, this species is predatory. It typically hides in vegetation and darts out to seize its prey. Newly hatched and young pickerel feed on plankton and small invertebrates. The adult fish feed on a variety of smaller fishes, amphibians, crustaceans, and the larvae of larger insects. The redfin pickerel primarily feeds during the day (55).

In the southern part of its range the redfin pickerel probably spawns from October to February in shallow water with heavy vegetation (55). Pickerel do not construct nests. Several hundred eggs are broadcast in and around submerged vegetation, to which the adhesive eggs cling. Eggs are not attended by the parent fish. Hatching occurs in approximately 10 days. Growth in the first 2 years is rapid—3–4 inches per year. Females are usually larger and live longer. The life span ranges from 3 to 7 years (61).

Comments. Two subspecies of *E. americanus* are recognized (21, 22). An Atlantic coastal form, *E. a. americanus* Gmelin (redfin pickerel), ranges from the St. Lawrence south to the St. Marys River, including the Okefenokee Swamp. A Mississippi Valley form, *E. a. vermiculatus* Lesueur (grass pickerel), ranges from the Pearl River of Louisiana west to the Brazos River of Texas. A considerable range of intergradation between the two forms occurs across Florida, western Georgia, and through Alabama and Mississippi. Intergrades inhabit portions of the Suwannee River from the Withlacoochee River south.

Sportsmen greatly enjoy taking all members of the pike family, but the redfin pickerel is less desirable because of its comparatively smaller size.

41

Chain Pickerel

Esox niger Lesueur

Other Local Names. Black pike, green pickerel, green pike, jack, jack-fish.

Scientific Name. *Esox*, from the Latin for "pike"; *niger* from the Latin meaning "dark" or "black."

Description. Larger than the redfin pickerel, adult chain pickerel range between 15 and 18 inches in length, attaining a maximum of 31 inches—the angling record, caught in Homerville, Georgia. The chain pickerel's body is elongate, cylindrical, and only slightly compressed laterally. The cheek and opercle of the elongate head are more or less scaled. The snout is duckbilled and longer than that of the redfin pickerel. The distance from the tip of the snout to the center of the eye is equal or greater than the distance from the center of the eye to the posterior margin of the opercle. Branchiostegal rays in the membrane along the lower margin of the gill cover numbers 14–16 on either side. The dorsal fin is located posterior, opposite the anal fin. The caudal fin is forked. There are 110–38 scales in the lateral line.

The back and sides are olive green to brown with a chainlike network of markings. These chainlike markings are not usually developed in young specimens of 6–8 inches in length. The ventral surface is creamy white. A dusky bar extends vertically downward from the eye. The dorsal and anal fins have dark pigment on the rays. The paired fins have less pigment.

Similar Species. The only similar species occurring in the Okefenokee Swamp is the redfin pickerel, *E. americanus.* The two species can be readily distinguished by a comparison of the snout length, the dusky bar beneath the eye, and the general pattern of markings. (See previous account for *E. americanus.*)

Distribution. The chain pickerel is found in most Atlantic and Gulf Coast drainages from Nova Scotia south to Florida and west to the Mississippi River. It occurs in both the St. Marys and Suwannee rivers, including the Okefenokee Swamp.

Habitat. This species prefers the quiet waters with thick plant growth of rivers and larger streams, lakes, ponds, and marshes. In the Okefenokee Swamp the chain pickerel inhabits canals, boat runs, prairies, lakes, and larger streams. It does not appear to penetrate smaller headwater streams and thus occurs in fewer areas of the swamp watershed than does the redfin pickerel. Studies in the East Prong Spillway of the Suwannee River indicate that chain pickerel make up approximately 2 percent of the fish caught (37). Similar figures were obtained for other locations in the swamp and along the upper portions of the Suwannee River.

Biology. Chain pickerel are solitary fish. They generally prefer to remain hidden in thick vegetation. Larger fish move to deeper waters during the day but tend to return to shallow waters at night. They are believed to establish territories or stations in summer, rarely leaving their areas except in the pursuit of food (54, 89). The chain pickerel is predatory. The newly hatched young feed on plankton and later shift to larger aquatic invertebrates. Larger chain pickerel feed more exclusively on fishes than do redfin pickerel. The diet also includes crayfish, large invertebrates as well as frogs, reptiles, and even birds and mammals (54, 89).

In the Okefenokee Swamp this species spawns primarily in the winter months of December and January and possibly as late as March and April. It breeds in the thick vegetation of shallow backwaters. The males attend the females, and both eggs and sperm are shed simultaneously. The eggs adhere to vegetation and hatch in 10–12 days. Growth is extremely rapid, with individuals reaching up to 8 inches in the first year. Females generally grow faster, reach sexual maturity earlier, and live longer than males. Both sexes may become sexually mature in the first year. The life span of the chain pickerel averages 3–4 years, with a maximum of 8–9 years (54, 89).

Comments. The chain pickerel is an excellent sport fish because of its vigorous fighting when hooked and the readiness with which it takes lures and live bait.

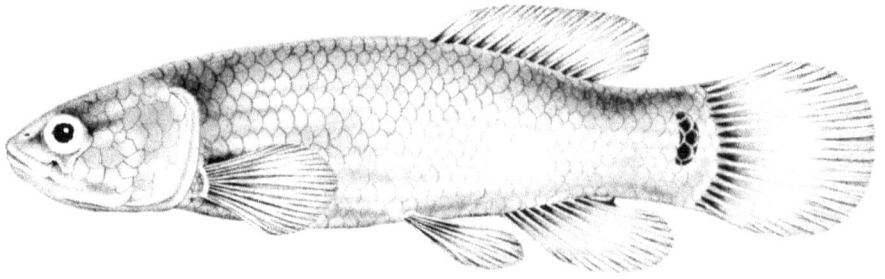

Eastern Mudminnow

Umbra pygmaea (DeKay)

Other Local Names.   Dog-fish.

Scientific Name.   *Umbra,* from the Latin for "shade"; *pygmaea,* from the Greek for "small."

Description.   The eastern mudminnow is a small fish, reaching a maximum size of approximately 3½–4 inches. The body is ovate, almost round in cross section. The head is broad, and the short snout is equal in length to the diameter of the eye. The upper jaw is nonprotractile. The cheeks, opercle, and body are covered with large cycloid scales. There is a single, soft-rayed dorsal fin, usually with 14 rays. The caudal fin is rounded. The pelvic and anal fins are abdominal in position and are located beneath the dorsal fin.

The top of the head, back, and sides are olive to brownish. The lower jaw is dark. There are pale, narrow, longitudinal stripes running the length of the body. At the base of the caudal fin is a distinct, thick (1½ scales), darkly pigmented vertical bar.

Similar Species.   The eastern mudminnow could be confused with a number of species—juvenile bowfin, the topminnows, the killifishes, the mosquitofish, or the pirate perch. The topminnows, killifish, and mosquitofish, however, all have a relatively deep, wide groove separating the tip of the upper lip and the snout while the eastern mudminnow does not. The pirate perch is easily distinguished by the presence of the anus in its throat region. The eastern mudminnow should not be easily confused with the bowfin because of the latter's long dorsal fin or with young pickerel, which have a forked caudal fin.

Distribution.   The species is confined to the Atlantic Coastal Plain from the Hudson River south to the St. Johns River in Florida and west along the Gulf Coast to the Aucilla River. It occurs in both the St. Marys and Suwannee rivers.

Habitat.   The eastern mudminnow prefers quiet waters with thick plant growth, particularly backwater ponds and pools of small creeks and streams that have heavily silted or muddy bottoms. In the Okefenokee Swamp it has been taken in small creeks and in the tributaries of larger streams.

Biology.   The eastern mudminnow is carnivorous and is primarily a bottom feeder. The diet consists largely of crustaceans, but insect larvae and other invertebrates are also consumed. It rarely feeds on other fishes (66).

Available evidence indicates that the eastern mudminnow breeds during late winter in southern Georgia and northern Florida. Spawning apparently occurs in the heavy vegetation of shallow backwater pools. No specific breeding information is available for this species; presumably it is similar to the closely related central mudminnow (*U. limi*), which lives in the upper Mississippi River and the Great Lakes region. In the latter species between 220–2,200 eggs are released by the female. They are adhesive and stick to nearby vegetation. No parental care is provided the eggs or young. The eggs hatch in one week. The female central mudminnow grows faster than the male and may live longer. Both become sexually mature by the second year. Maximum longevity is around 4 years.

Comments.   The eastern mudminnow frequently burrows into the mud, tail first, when frightened or disturbed. It also apparently has the ability to burrow into moist mud if the water in which it lives dries up (30). This is possible because its swim bladder is physostomous, which enables the fish to utilize atmospheric oxygen.

When the fishes of the Okefenokee Swamp were first investigated seriously by Palmer and Wright (64), their report of the occurrence of a mudminnow in the swamp was a significant find. The fish had never before been found in this region of the Southeast. Palmer and Wright, however, identified the species occurring in the swamp to be *U. limi* and not *U. pygmaea*. Some problems with the taxonomy of the species remain because specimens collected in Baker County, Florida, to the south of the swamp have been referred to as *pygmaea* but have a higher dorsal and anal fin ray count than either *pygmaea* or *limi*.

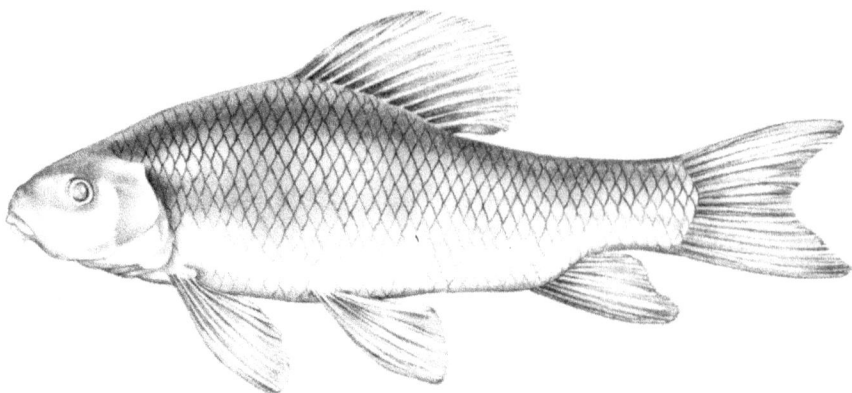

Lake Chubsucker

Erimyzon sucetta (Lacepède)

Other Local Names. Mullet, creek fish, sucker.

Scientific Name. *Erymizon,* from the Greek meaning "to suck"; *sucetta,* from the French referring to "a sucker."

Description. This is a small "sucker," usually only 5–7 inches in length and rarely exceeding 10 inches. It is a chubby, rather deep-bodied fish with a strongly arched back. The species lacks a lateral line. There are generally 35–38 lateral scales. The head is bluntly rounded and the snout extends only slightly beyond the upper lip of the protrusible mouth. The dorsal fin is short with 11–12 fin rays. The posterior margin of this fin is convex. The caudal fin is shallowly forked and the tips are rounded. The anal fin is prominent. Breeding males have a distinctive bilobed anal fin and tubercles generally appear on either side of the snout.

The back and upper sides range in color from greenish bronze to olive brown. The scales lack dark spots but are distinctly dark-edged. The dark lateral stripe along the sides may not be well developed in adults but is usually distinct in the young. The dorsal and tail fins are olive- to slate-colored and the ventral fins are whitish yellow. Breeding males are not distinctively colored.

46

Similar Species. The lake chubsucker is not easily confused with other fishes, though it may be confused with certain minnows. True minnows are not known to occur in the acidic waters of the Okefenokee Swamp, yet they do occur further downstream in both the Suwannee and St. Marys rivers. The lake chubsucker is easily distinguished from the spotted sucker, which has distinct horizontal rows of dark spots. Unlike the chubsucker, the spotted sucker has a concave dorsal fin and 43–45 midlateral scales.

Distribution. The lake chubsucker is restricted to eastern North America. It is found throughout much of the lower Great Lakes Basin, south through the Mississippi Valley, along the Gulf Coast from Texas to Florida—including the St. Marys and Suwannee rivers—and north along the Atlantic Coastal Plain to Virginia.

Habitat. This species inhabits the quiet waters of ponds, lakes, canals, prairies, sloughs, creeks, and borrow pits. It prefers sandy to muddy bottoms with much organic debris in areas of both dense submerged and emergent vegetation. Generally, the young are found more frequently in dense vegetation around the water's edge; adult fish live in deeper open water.

Biology. Like other suckers, this species is a bottom feeder. Young lake chubsuckers feed largely on plankton. Adult fish are omnivorous, feeding on small crustaceans such as water fleas and larval aquatic insects. To a lesser extent, algae and other plant matter are also eaten (55).

Spawning lasts approximately two weeks, and available evidence in the Southeast indicates that it may take place over a relatively long period, occurring as early as December or as late as May (55). After a brief courtship, the female broadcasts from 2,000 to 20,000 nonadhesive eggs in and around vegetation. The eggs hatch in about one week. Hatchlings are scarcely over ¼ inch, but the young may reach 2 inches the first year and will grow approximately 2 inches more per year. Average longevity is 4–5 years, with 8 years as a maximum.

Comments. This species is generally too small to be considered a popular game fish. Because of their small size, they are rarely eaten, although they do taste good. Like other suckers, chubsuckers are not often taken on hook and line; they are usually snagged or gigged. In some areas the chubsucker has been transplanted to serve as a forage fish for largemouth bass. The distinction of an eastern and western subspecies was proposed earlier in this century (41), but this view is unsupported by more recent authorities (4).

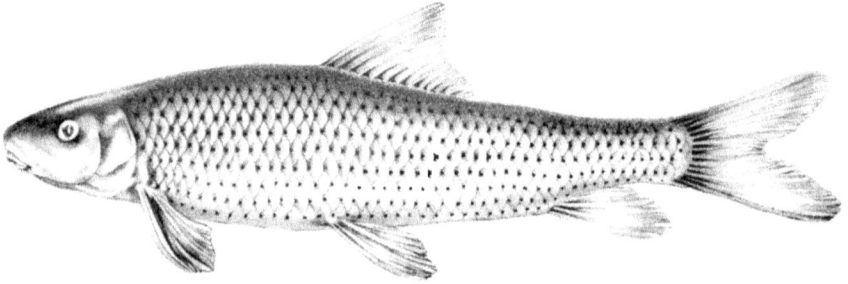

Spotted Sucker

Minytrema melanops (Rafinesque)

Other Local Names.   Sucker.

Scientific Name.   *Minytrema*, from two Greek words meaning "reduced aperture," in reference to the poorly developed lateral line; *melanops* from the Greek referring to its "black spotted" appearance.

Description.   This is a medium- to large-sized fish, ranging 9–16 inches in length, infrequently attaining a maximum of 22 inches. Its body is moderately elongate and slender. The head is bluntly rounded. The protrusible mouth is subterminal with the snout extending well beyond the upper lip. A lateral line is absent or poorly developed. There are 43–45 midlateral scales. The dorsal fin has 11–12 rays, and its posterior margin is slightly concave. The caudal fin is deeply forked. Breeding males have small tubercles on the snout, cheek, and behind the eye.

   The back of the head and body are dark olive brown, the sides brassy or silvery, and the belly whitish. The common name results from the presence of dark spots on the base of the scales on the sides. These form distinct horizontal rows of spots that are usually absent in young specimens less than 3 inches. The dorsal and caudal fins are gray to olive and the paired fins are dusky white. Breeding males typically have two dark longitudinal bars separated by a grayish pink band along the sides.

Similar Species. The spotted sucker, because of the distinct rows of dark spots, is easily distinguished from the lake chubsucker. The lake chubsucker also has fewer midlateral scales (35–38) and has a convex dorsal fin.

Distribution. The spotted sucker ranges throughout the lower Great Lakes Basin and south through the Mississippi Valley. It occurs in Gulf Coast drainages from Texas east to the Suwannee River. It is not known from peninsular Florida east of the Suwannee River or the St. Marys River, although it occurs to the north along the Atlantic Coastal Plain as far as North Carolina.

Habitat. The spotted sucker usually lives in sluggish waters with heavy vegetation and a fairly solid substrate containing organic debris. It is known from overflow pools of rivers and streams, oxbows, and sloughs. There are no records of this species living within the swamp refuge boundaries; however, it does occur in appropriate habitats along the Suwannee River north of Fargo as well as in some of the smaller tributaries of the Suwannee River within the Okefenokee Swamp watershed boundary. The spotted sucker is considered to be intolerant of turbid and/or polluted water. As such, it is a good indicator species of water quality.

Biology. Like other suckers, the spotted sucker is a bottom feeder. It is reported to consume large quantities of detritus (7, 88), although it also feeds on small crustaceans, the larvae of aquatic insects, and algae.

In the southeastern United States the species has been reported to breed from mid-March until early May (55, 56). It spawns in shallow riffles of small tributary creeks. Eggs hatch in 1–2 weeks, depending upon water temperature. The young grow rapidly and may attain a length of 6 inches in the first year and approximately 11, 13, 16, and 17 inches in succeeding years. The fish matures in 3 years and the maximum life span is usually 5–6 years (44).

Comments. The spotted sucker is a popular game and food fish in most areas of the South. It is most frequently taken by gigging. Local residents value it highly, though this opinion is not shared by all. David Starr Jordan, a pioneer in American ichthyology during the last century, commented that the fish was "pretty good for a sucker, which is not saying much." Joshua Laerm insists that this is one of the best fish for braising (simmering in fluids) and suggests cooking the fish slowly in white wine with carrots and onions.

Yellow Bullhead

Ictalurus natalis (Lesueur)

Other Local Names. Cat, catfish, mud cat, yellow cat.

Scientific Name. Ictalurus, from the Greek meaning "fish-cat"; *natalis,* from the Latin for "buttock" or "rump."

Description. The yellow bullhead is a medium-sized but stout-bodied catfish usually 8–12 inches in length and reaching a maximum of 18 inches. The head is broad, with 4 pairs of sensory barbels, the longest reaching to the base of the pectoral fin. Sharp, robust spines occur on the pectoral and dorsal fins. The posterior margin of the pectoral spine bears small toothlike serrations. An adipose fin is present; it is short and its posterior margin is free and well separated from the caudal fin. The rear margin of the caudal fin is nearly straight (in young) to slightly rounded (in adults). The yellow bullhead has 23–27 anal fin rays and no scales.

The top of the head, back, and upper sides range from yellowish brown to dark olive brown. The sides are yellow to yellow-brown, and the belly, yellow to white. The upper barbels are light to dark brown, but the chin barbels are uniformly whitish with no black pigment or spots. The anal fin may have a vague, broad, horizontal dark band along its center.

50

Similar Species.   Catfish are not easily confused with other groups of fishes because of their distinctive barbels and scaleless bodies. The yellow bullhead is easily distinguished from the brown bullhead, which has dark barbels, the channel catfish, which has a deeply forked caudal fin, and the madtom, whose adipose fin is continuous with the caudal fin.

Distribution.   The yellow bullhead occurs widely throughout the eastern and central United States and ranges into extreme northeastern Mexico and Ontario, Canada. It is common throughout the St. Marys and Suwannee rivers and the Okefenokee Swamp.

Habitat.   This species prefers living in the heavy vegetation found in the littoral areas of relatively clean waters with a permanent flow. It generally avoids rapid current or deeper open waters. In the Okefenokee Swamp it has been taken from lakes, canals, and prairies, as well as from the backwater pools of rivers, streams, and creeks. In the upper Suwannee River and many areas of the swamp the yellow bullhead is the most commonly caught fish (39).

Biology.   The yellow bullhead is a nocturnal bottom feeder that searches for food with its sensory barbels. Adult fish are omnivorous and include primarily crustaceans, aquatic insects, molluscs (snails), and to a smaller degree small fishes and even some aquatic vegetation in their diet.

The yellow bullhead breeds primarily in late spring and early summer, but breeding may occur during all months of the year (55). One or both parent fish prepare a shallow depression for a nest, usually under some cover. Several thousand eggs are deposited in a mass in the nest, which one or both parent fish guard. The eggs hatch in 5–10 days. The young aggregate and are not abandoned by the parent fish until they are about 1 inch in length. The young continue to school throughout their first summer. Growth is rapid, averaging more than 7 inches the first year (81). The yellow bullhead reaches sexual maturity in 2–3 years and has a maximum life span of 6–7 years.

Comments.   The yellow bullhead is an excellent sport fish and is an important component of creel records in the Okefenokee Swamp. Almost any kind of bait may be used. In fact, quite a varied folklore exists on the best "cat" bait, ranging from corn kernels and snails (apparently a preferred natural food) to rancid cheese and rotten liver. The yellow bullhead is an excellent food fish, particularly tasty when fried, although the skin is thin and some difficulty may be encountered in "dressing" the fish.

51

Brown Bullhead

Ictalurus nebulosus (Lesueur)

Other Local Names. Brown cat, cat, catfish, mud cat, red cat.

Scientific Name. *Ictalurus*, from the Greek meaning "fish-cat"; *nebulosus*, from the Latin meaning "misty," "foggy," or "cloudy," in reference to its mottled appearance.

Description. The brown bullhead is a medium-sized catfish usually 7–15 inches in length and reaching a maximum of 21 inches. This species is less robust than the yellow bullhead. The head is massive. The longest barbel reaches well beyond the base of the pectoral fin. Sharp spines are prominent on the dorsal and pectoral fins. The rear margin of the pectoral spine is strongly serrated. As in the yellow bullhead, the adipose fin is short with its rear margin free and well separated from the caudal fin. The rear margin of the caudal fin is slightly notched. The fish has 21–24 (usually 22–23) anal fin rays.

The top of the head, back, and upper sides range from yellow-brown to olive to almost black. The sides are usually strongly mottled with darker brown or black. The belly is yellow to dirty white. All barbels are brown to black—except sometimes the base of the chin barbels may be lighter. The fins are dusky with the rays darker than the membranes between them.

Similar Species.   The brown bullhead is easily distinguished from other catfish: the yellow bullhead, which has whitish chin barbels; the channel catfish, which has a deeply notched caudal fin; and the two madtoms, whose adipose fins are continuous with their caudal fins.

Distribution.   This species occurs throughout most of the eastern and north central United States and adjacent southern Canada. It is found in both the St. Marys and Suwannee rivers.

Habitat.   The brown bullhead is most often found in heavy vegetation of warm water having little or no current and a silty or muddy bottom with accumulated debris. It prefers the slow-moving backwaters of rivers and streams, ponds, and the bays of large lakes.

The brown bullhead is not recorded to live within the boundaries of the Okefenokee National Wildlife Refuge. It does, however, occur in small tributary creeks of the Suwannee River.

Biology.   Like other catfish, the brown bullhead is a nocturnal bottom feeder. Its diet consists of almost any available resource, including crustaceans, aquatic insects, and molluscs. It will feed on smaller fishes and fish eggs as well as on some plant material, including blue-green algae (70).

Breeding generally occurs in the spring and early summer but may extend throughout the year, particularly in the South, where the brown bullhead sometimes spawns twice a year. Either one or both parent fish prepare a shallow nest, usually in or near submerged vegetation or debris. As many as 13,000 adhesive eggs may be released. The eggs are cared for by one or both parent fish. Eggs hatch in 6–9 days. The young school and are guarded by one or both parent fish until they are approximately 2 inches in length, at which time the young disperse. Growth is moderately rapid. Sexual maturity is reached by the third year. The life span is 6–8 years (11, 12).

Comments.   Like other catfish, the brown bullhead is an excellent sport fish. As Henry David Thoreau wrote, "They will take any kind of bait, from an angle worm to a piece of a tomato can, without hesitation or coquetry, and they seldom fail to swallow the hook." Many local residents, however, feel the food quality of this fish is inferior to other catfish.

Two subspecies are recognized, *I. n. nebulosus* north of Virginia and *I. n. marmoratus* to the south.

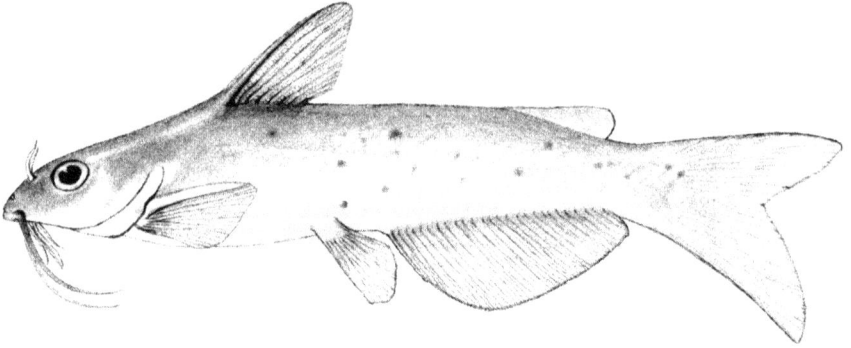

Channel Catfish

Ictalurus punctatus (Rafinesque)

Other Local Names.   Cat, catfish, spotted catfish.

Scientific Name.   *Ictalurus,* from the Greek meaning "fish-cat"; *punctatus,* from the Latin, in reference to its spotted appearance.

Description.   The channel catfish is the largest of the catfish in the swamp. Adult fish are usually 12–21 inches in length with the rare specimen attaining 36 or more inches. The body is long and slender. The head is massive, with the upper jaw protruding beyond the lower jaw. Dorsal and pectoral spines are present. The pectoral spine lacks the serrations found in the yellow bullhead and the brown bullhead. The adipose fin is short with its rear margin free and separated from the caudal fin. Unlike the other catfish, the caudal fin is deeply forked. The channel catfish has 24–29 anal rays.

The top of the head, back, and upper sides are pale blue to a silvery brownish olive. The lower sides are lighter and the belly is silvery white. The color of the fins is similar to the adjacent body parts. The barbels are dusky-colored without spots. This is the only spotted catfish that has irregularly arranged dark spots of varying size along the sides. Breeding males may be deep blue-black on the back and upper sides.

Similar Species.   The channel catfish is the most distinctive of all the catfish of this region because of its spots and deeply forked tail.

Distribution. The natural distribution of this species is central and eastern United States and adjacent regions of northeastern Mexico and southern Canada. It does not, however, occur in the coastal plain of the mid-Atlantic states or in New England. The species has been widely introduced outside the natural range, particularly in the southwestern United States.

Habitat. The channel catfish occurs primarily in larger streams and rivers and their connecting lakes and backwaters. During the day the species prefers deep water, where the fish generally remain under logs, rocks, vegetation, and among other bottom debris that provide suitable cover. Although the channel catfish is very common in the St. Marys and Suwannee rivers and their major tributaries in the Okefenokee Swamp watershed, there is no record of its being taken in the Okefenokee National Wildlife Refuge.

Biology. Like other catfish, the channel catfish is most active at night, when it moves into shallow pools and riffles to feed. Its diet is extremely varied and includes detritus (partially decomposed organic matter), crustaceans, molluscs, aquatic insects, small fishes, algae, and other aquatic plant material.

In the South this species spawns from May to September (55). Unlike some other catfish, female channel catfish do not participate in nest site selection or care of the young. Nest sites include almost any semidark or secluded spot, such as undercut banks, submerged debris, or hollow logs, or even burrows of muskrats or beavers. More than 35,000 eggs may be released into the nest and are attended by the male. The eggs hatch in one week and the male guards the fry until they leave the nest, generally within a week. Growth is rapid, between 2 to 4 inches the first year. Sexual maturity occurs in 4–5 years, and longevity averages 6–7 years.

Comments. The channel catfish is doubtless the most economically important catfish in North America. It is widely propagated as a pond culture fish, and there is an extensive literature dealing with its fishery (60).

The channel catfish is one of the most popular sport catfish in most areas of its range. It will take nearly any live, cut, or prepared bait, and even artificial lures or spinners. It is a favorite food fish of most southerners, and its flavor is unsurpassed by that of any other catfish.

Tadpole Madtom

Noturus gyrinus (Mitchill)

Other Local Names. Madtom, tadpole cat.

Scientific Names. *Noturus*, from two Greek words meaning "back tail," in reference to the connection between the adipose and caudal fins; *gyrinus*, from the Greek for "tadpole."

Description. This is a very small catfish, usually 2–3 inches in length and reaching a maximum of 4 inches. The scaleless body resembles that of a tadpole, hence its common name. The head is massive, the anterior trunk, potbellied and tapering posteriorly. The upper and lower jaws are of equal length. Dorsal and pectoral spines are present. The pectoral spine is long and stout but without serrations. It has, however, a deeply grooved, posterior margin. The adipose fin is long and fleshy and separated from the caudal fin by only a shallow notch. The caudal fin is broadly rounded.

The back and sides are tan to brown, the sides are lighter, and the belly is pale yellow to whitish. The fins and barbels are the same color as adjacent body parts and are not freckled with dark spots. A dark, veinlike streak is present along the midlateral line.

Similar Species. The madtom is easily distinguished from other catfish because its adipose fin is continuous with the caudal fin or separated from it by only a shallow notch. The tadpole madtom should not be confused with the freckled madtom because of the distinct freckling in the latter.

Distribution. Limited to eastern North America, the tadpole madtom occurs throughout the lower Great Lakes Basin and Mississippi Valley into the Gulf Coast region, where it ranges from Texas to Florida, including the St. Marys and Suwannee rivers. Its range extends north along the Atlantic Coastal Plain into New England. It does not, however, occur in the Appalachian Highlands.

Habitat. The tadpole madtom inhabits moderately sluggish to slow-moving waters. It occurs in creeks and small streams and their backwaters, preferring areas with thick vegetation and accumulated organic debris. In the Okefenokee Swamp it has been caught from canals, boat runs, lakes, streams, creeks, and sloughs.

Biology. Like most catfish, the tadpole madtom is secretive during the day, retiring to areas of thick plant growth or submerged debris. It feeds on the bottom at night, eating primarily planktonic crustaceans, immature aquatic insects, and rarely small fish.

Although in most areas of its range it breeds in late spring or early summer, it may breed in late winter or early spring in the Okefenokee Swamp. Compared to other catfish, the tadpole madtom releases relatively few eggs, usually between 50 and 125. Nests are normally prepared in dark cavities within or under submerged debris—even in tin cans. The adhesive egg clusters are surrounded by a gelatinous envelope and may be attended by one of the parent fish. The young generally attain a length of 2 inches or more during the first year. Most fish become sexually mature during their second year. Maximum longevity is 2–3 years.

Comments. This catfish is too small to be of interest to sportsmen, although it may be used as an effective bait for bass. Both the dorsal and pectoral spines are associated with a poison gland located at their respective bases (71). These glands are ductless and poison is only released when the epidermis of the spine is disturbed. The wound from these spines can be painful to humans, but generally less so than that of a bee sting. The poison gland does not appear to bother the catfish's predators however.

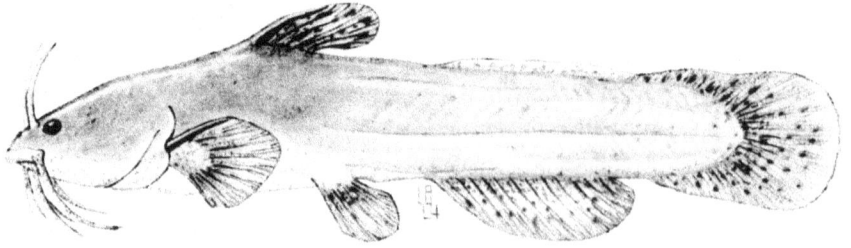

Speckled Madtom

Noturus leptacanthus Jordan

Other Local Names.   Gulf madtom, madtom.

Scientific Name.   *Noturus*, from two Greek words meaning "back tail," in reference to the connection between the adipose and caudal fins; *leptacanthus*, from two Greek words meaning "slender spine."

Description.   The speckled madtom is a very small catfish; adults average 2–3 inches, attaining a maximum of approximately 3½ inches. The body is slender and elongate; its standard length is more than 4.5 times its maximum depth. The head is small and narrow, more long than broad. The upper jaw projects beyond the lower jaw. The dorsal and pectoral fins are small and short. The fin spines are short and slender with slight serrations on the anterior edge. The adipose fin is continuous with the rounded caudal fin.

The top of the head, back, and upper sides are a mottled, reddish to brownish yellow. The belly is whitish yellow to white. The barbels are colored similarly to adjacent body parts—dusky above, light below. Body and fins are speckled with darker pigment.

Similar Species.   The speckled madtom can easily be distinguished from the tadpole madtom because the latter lacks speckling on the body and fins. Madtoms are easily distinguished from other catfish because their adipose and caudal fins are continuous.

Distribution.   This species is limited to the southern Atlantic and Gulf slope drainages from the Edisto River, South Carolina, to the Amite and Comite rivers in Louisiana. It occurs in both the St. Marys and Suwannee rivers.

Habitat.   The speckled madtom inhabits shallow waters with thick plant growth in creeks, rivers, and their smaller tributary streams. The fish is apparently tolerant of moderate current and may be found in riffle areas, particularly in smaller creeks. It is most often found near submerged debris and/or areas of dense vegetation and is strictly a bottom dweller. This species has not been recorded in the Okefenokee National Wildlife Refuge. It does, however, occur in the Okefenokee Swamp watershed in the Suwannee River and its tributary creeks above Fargo.

Biology.   Very little is known of the feeding and reproductive biology of this species. Like other catfish, it apparently tends to be secretive during the day, seeking refuge in and around submerged debris. The speckled madtom feeds primarily at night and—like the tadpole madtom—includes crustaceans as well as aquatic insects in its diet.

The range of the spawning season is not known. Records do exist of spawning from July to August (17), but indirect evidence suggests that spawning may extend from April to October and possibly later (55). Some females may breed twice in one season. It appears that this species is among the least fecund of egg-laying fish. Only 15–30 eggs are known to be deposited. One or both parent fish prepare a nest in or around submerged debris—frequently old cans or bottles. The male guards the eggs and larvae. The eggs hatch after approximately one week. The young become sexually mature and breed the next year, but apparently few survive for a second breeding season.

Comments.   This species is obviously too small to be of interest as a food or game species, although it could serve as bait for bass.

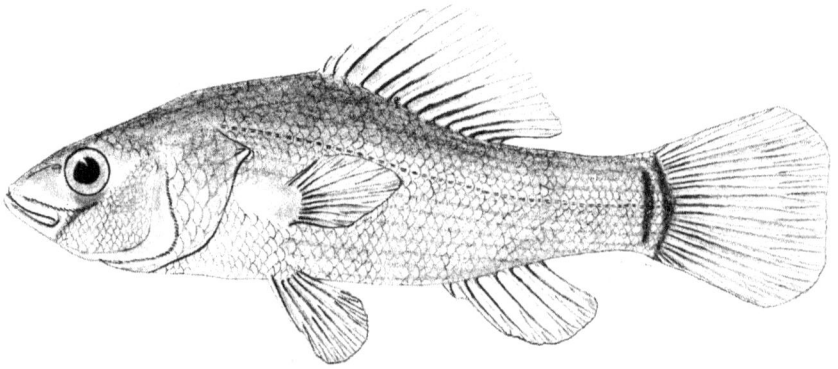

Pirate Perch

Aphredoderus sayanus (Gilliams)

Other Local Names. Asshole perch, politician fish.

Scientific Name. *Aphredoderus,* from the Greek words "aphodos" meaning "excrement" and "dere" meaning "neck," in reference to the position of the anus in the throat region; *sayanus,* named after Thomas Say, an entomologist.

Description. The pirate perch is a stout, deep-bodied, small fish ranging in size from 2½ inches to a maximum of 5½ inches. It has a large mouth. Each opercle has a sharp spine on its rear margin, and the rear margin of the preopercle is strongly serrated. There is a single dorsal fin and the tail is slightly notched. The pirate perch differs from other fish in the area by having its anus located far forward; in adult fish the anus is in the throat region.

The head and body appear dark olive to grayish black as a result of a profuse speckling. The underside of the head and belly sometimes is yellowish. A dark vertical bar is prominent beneath the eye. There is another narrow, dark vertical bar at the base of the caudal fin. The median fins (dorsal and anal) have a deeply pigmented base and narrow, marginal white fringe. Breeding males and females are quite iridescent, predominantly violet and purple. Breeding males may appear almost black.

Similar Species. The pirate perch may superficially resemble bream or bass, but the latter two groups possess two dorsal fins.

60

Distribution. This species occurs throughout the lowlands of the Atlantic and Gulf coasts from New England to Texas and up through the Mississippi Valley to the Great Lakes. It occurs in both the St. Marys and Suwannee rivers, and also in much of the Okefenokee Swamp.

Habitat. The pirate perch inhabits lowland lakes and overflow ponds and pools of slow-moving streams. It is usually found in waters with abundant vegetation. In the Okefenokee Swamp it inhabits canals, boat runs, lakes, prairies, and smaller tributary streams. It is not a common fish but may be locally abundant in appropriate habitat.

Biology. The pirate perch received its common name from Dr. C. C. Abbott because it habitually ate only other fish when it was in his aquarium. Normally the pirate perch feeds predominantly on the larvae of aquatic insects and small crustaceans (30). Individual fish are solitary and secretive. They spend the daylight hours in thick vegetation and usually emerge only at night to feed.

Very little is known of the reproductive habits of the pirate perch. Available evidence in Georgia and Florida suggests it breeds in early winter, October through December. It has been suggested that the pirate perch builds a nest that is guarded by both parent fish for sometime after the eggs hatch (26, 30). There is also some speculation that the eggs may be incubated in the gill covers as occurs in closely related cave fishes (67). The young may grow 2½ inches the first year and 1 inch in succeeding years. Few survive longer than 4 years.

Comments. The pirate perch may occasionally be taken by sportsmen and confused with a small bream or bass; however, they are easily distinguished. Its small size generally precludes its being a popular food or game fish.

The development and position of the anus is very interesting in these fish. In young, the anus is located in front of the anal fin, but during growth the anus migrates anteriorly to the throat region.

There are two recognized subspecies of pirate perch. *Aphredoderus s. sayanus* occurs along the Atlantic slope south to Florida and including the Okefenokee Swamp. *Aphredoderus s. ibbosus* occurs throughout the Mississippi Valley. The precise range of the subspecies has not been determined, but it is suggested to be similar to that of *Esox americanus* (4, 51).

This quiet, but interesting fish is often popular in aquaria, where it sometimes rests in almost a vertical position with its head pointed down.

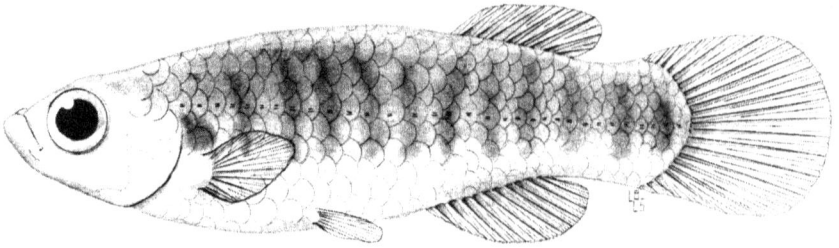

Male

Golden Topminnow

Fundulus chrysotus (Günther)

Other Local Names.   Killifish, minnow, redfin minnow, topminnow.

Scientific Name.   *Fundulus,* from the Latin for "bottom," the abode of the *Fundulus* mudfish; *chrysotus,* from the Greek for "gilded" or "golden."

Description.   The golden topminnow is a small fish rarely exceeding 2 inches in length. The body is relatively elongate and stout. The head is short, wide, and flat. The snout is also short (about two-thirds the diameter of the eye) as is the mouth. The origin of the dorsal fin is opposite the third ray of the anal fin. There are 7–9 dorsal fin rays and 9–11 anal rays. The caudal fin is bluntly rounded.

The back and upper sides of the golden topminnow are dark yellow-green to olive. There is a small golden spot near each nostril and a short, narrow golden stripe in front of the dorsal fin. Males and females exhibit considerable sexual dimorphism. The females and the young of both sexes have scattered, golden or pearly spots on the sides, particularly on the posterior part of the body. In males the spots are reddish. The median fins of males are reddish (hence their common name) with darker, scattered reddish spots. Males also have 10–12 broad, wavy vertical bars whose edges are somewhat indistinct. Females lack these bars.

Similar Species.   This species most closely resembles, and is often confused with, the banded topminnow. The latter, however, lacks golden or pearly spots

62

on the sides and has an elongate golden spot above and behind the eye. While both species have vertical bars, the banded topminnow has more, usually in excess of 12, and the bars are narrow, straight, and have distinct edges. The golden topminnow is easily distinguished from the lined topminnow, which has a distinct, dark, teardrop-shaped blotch beneath the eye, and the pygmy kill-ifish, which has a large, distinct dark spot at the base of the caudal fin. Super-ficially, the golden topminnow resembles several other small fish: the eastern mudminnow, which has a larger number of dorsal fin rays, usually 14; the least killifish, which has a distinct, dark horizontal bar and pronounced dark spots on the dorsal and caudal fins; the pirate perch, whose caudal fin is deeply notched and whose anus is located in the throat region; and the mosquitofish, in which the origin of the dorsal fin is over or behind the posteriormost anal fin ray.

Distribution.   The species occurs in the lowland Atlantic Coastal Plain from South Carolina throughout Florida and west along the Gulf Coast to eastern Texas. It also ranges up the Mississippi River basin to extreme southern Mis-souri and Kentucky.

Habitat.   The golden topminnow inhabits clear, quiet backwaters and pools of slow-moving lowland rivers and streams. It is not commonly found in brackish water. It usually prefers waters with submerged vegetation. Although the golden topminnow is known from both the St. Marys and Suwannee rivers, it is not widely distributed in the Okefenokee Swamp, having been recorded only from prairies, canals, and a few tributary streams. It is usually collected under mats of floating vegetation.

Biology.   This fish feeds almost exclusively at the water's surface. Its diet con-sists primarily of small crustaceans and aquatic insects (43, 55).

No breeding information is available for this fish in the Okefenokee Swamp; in nearby Florida, however, the species apparently breeds from May until July. During breeding, the male attends the female in a courtship behavior. A single egg is released by the female and is fertilized by the male, who swims alongside. This behavior is repeated 10–20 times per day for a week or more. The eggs are deposited in rocks or in aquatic vegetation and hatch in 10–14 days (52).

Comments.   This small, secretive species is of little commercial or sport value to man. It is a good aquarium fish.

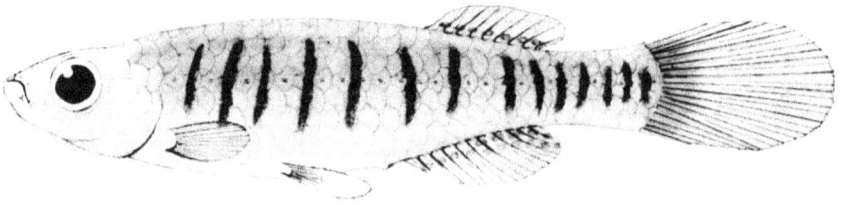

Male

Banded Topminnow

Fundulus cingulatus Valenciennes

Other Local Names.   Minnow, topminnow.

Scientific Name.   *Fundulus*, from the Latin for "bottom," the abode of the *Fundulus* mudfish; *cingulatus*, from the Latin for "belted," in reference to the vertical bands of pigment.

Description.   This small fish rarely exceeds 2 inches. Its body is relatively elongate, robust, and somewhat laterally compressed posteriorly. The head is small and flat and broad between the eyes. The small mouth is terminal. The origin of the dorsal fin is nearly opposite that of the anal fin. There are usually 9 dorsal rays and 10 anal rays. The caudal fin is rounded.

The back and upper sides are olivaceous. The scales on the side may be edged with dusky pigment forming a few faint longitudinal stripes. Both males and females have 12 or more dark vertical bars that are narrow and straight and have distinct edges. Occasional specimens have irregular black blotches scattered dorsally. There is a narrow, golden middorsal line anterior to the dorsal fin and also an elongate golden spot above and behind the eye. Breeding males have deep red dorsal and anal fins bordered by a fine black margin.

Similar Species.   The banded topminnow most closely resembles the golden topminnow; however, the latter lacks the golden spot located near the eye. Both fish are vertically barred, but the bars of the golden topminnow are less than 12 in number and are wavy and indistinct. The lined topminnow can be distinguished by its dark, teardrop-shaped blotch beneath the eye. The pygmy killifish is similar in shape but has a large, distinct dark spot at the base of the caudal fin. The banded topminnow superficially resembles: the eastern mudminnow, which has a large number of dorsal fin rays, usually 14; the least killifish, which has a distinct, dark horizontal bar and pronounced dark spots on the dorsal and caudal fins; the pirate perch, whose caudal fin is deeply notched and whose anus is located in the throat region; and the mosquitofish, in which the origin of the dorsal fin is over or behind the posteriormost anal fin ray.

Distribution.   This species is almost entirely confined to Florida south as far as the Tamiami Canal and north in Georgia to the Satilla River. The range extends west to Mobile Bay in Alabama. The fish occurs commonly in both the St. Marys and Suwannee rivers and widely throughout the Okefenokee Swamp.

Habitat.   The banded topminnow is a surface dweller that inhabits backwaters of sluggish lowland rivers and streams. It is common in marshes and swamps, typically in marginal areas with thick plant growth. In the Okefenokee Swamp it occurs in most aquatic habitats, including roadside ditches and small, pine, flatwood ponds. It is particularly common in *Utricularia* mats.

Biology.   The diet of this small fish is restricted to small aquatic organisms, principally small crustaceans and aquatic insect larvae as well as filamentous algae. There are reports that the species may also feed occasionally on the bottom (31).

The breeding season extends from March to October. Almost no specific information on reproduction is available, although presumably it would be similar to other members of the genus. There are suggestions that eggs may often survive in dry substrates as occurs in some other species of the family (31).

Comments.   Like other *Fundulus* this species is too small to be of significant commercial value to man. It is, however, an excellent aquarium fish.

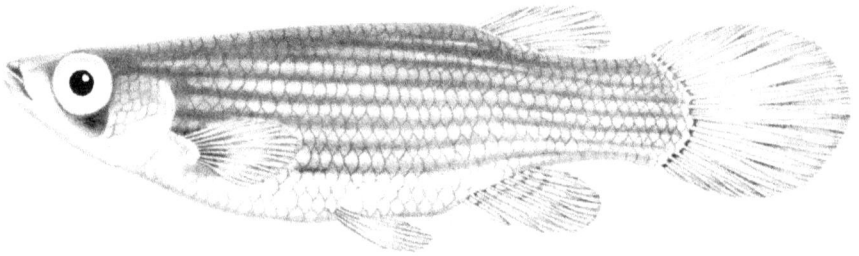

Female

Lined Topminnow

Fundulus lineolatus (Agassiz)

Other Local Names.   Minnow, star-head minnow, topminnow.

Scientific Name.   *Fundulus,* from the Latin for "bottom," the abode of the *Fundulus* mudfish; *lineolatus,* from the Latin referring to "lined."

Description.   The lined topminnow is small, usually 1–1½ inches. The body is relatively elongate and stout but somewhat compressed posteriorly. The head is broad and slightly concave on top. The mouth is terminal and small. The first ray of the dorsal fin is slightly behind the first anal fin ray. There are 7–8 dorsal rays and 8–10 anal rays. The caudal fin extends almost to a point rather than being bluntly rounded.

  The general ground color is silvery with dark body markings, some of which are different in males and females. Almost all individual fish have a dark teardrop below the eye and a dark pigment patch on the upper half of the pectoral fins. Both sexes usually have 7–8 horizontal stripes, which are typically darker and more distinct in females and are pale to absent in males. Males have 11–15 dark vertical bars and dark spots in the center of the scales on the sides of the body. These are usually absent in adult females. The best field character for this species is the distinct, light-colored, star-shaped blotch on the head.

Similar Species.   This topminnow can be distinguished from other topminnows by the absence of the characteristic teardrop beneath the eye. The pygmy killifish also lacks this teardrop and has a distinct spot on the base of the caudal fin. Superficially, the lined topminnow also resembles: the eastern mudminnow, which has a larger number of dorsal fin rays, usually 14; the least killifish, which has a distinct, dark horizontal bar and pronounced dark spots on the dorsal and caudal fins; the pirate perch, whose caudal fin is deeply notched and whose anus is located in the throat region; and the mosquitofish, in which the origin of the dorsal fin is over or behind the posteriormost anal fin ray.

Distribution. The species occurs along the Atlantic Slope from Virginia to Dade County, Florida, and from there west and north throughout Florida and southern Georgia to the Ochlockonee River.

Habitat. Throughout its range the lined topminnow is a fairly common inhabitant of quiet, clearwater streams, backwaters, and ponds. It is widely distributed and very abundant in the Okefenokee Swamp, occurring in nearly all aquatic habitats and especially in canals and lakes. It is also known from most creeks and the backwater areas of the drainage rivers.

Biology. The lined topminnow is probably more highly restricted to the surface of the water than other species of topminnows. It feeds primarily on small crustaceans and the larvae of aquatic insects.

Almost nothing is known about the breeding biology of the species. Presumably, however, its breeding habits are similar to other members of the genus. It probably breeds in early summer (36, 55).

Comments. The species is sufficiently abundant and so easily dipnetted that it is widely used as bait for game fish.

The lined topminnow is closely related to the starhead topminnow, *Fundulus notti*, which occurs from Mobile Bay to the Mississippi River. In fact, until recently the lined topminnow was considered a subspecies of *F. notti* (14, 74, 90).

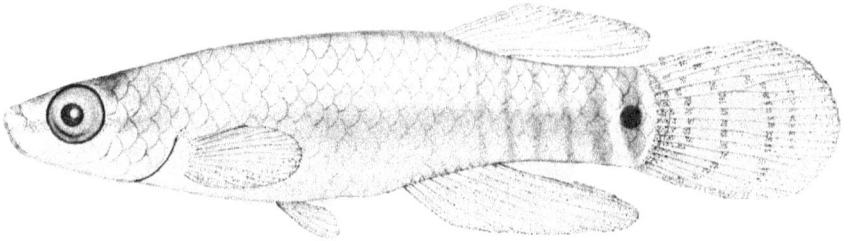

Male

Pygmy Killifish

Leptolucania ommata (Jordan)

Other Local Names.   Rainwater fish, target fish.

Scientific Name.   *Leptolucania,* the root "lepto" is from the Greek referring to small or delicate, but "lucania" is a coined word having no specific meaning, but possibly referring to the genus *Lucania; ommata,* from the Greek referring to "eyed," in reference to the spot at the base of the caudal fin.

Description.   This small, slender, fusiform fish does not exceed 1 inch. The mouth is small and the premaxillary bones very protractile. The lower jaw projects noticeably. The head appears small due to the large size of the eye, which is twice the length of the snout. There is a single dorsal fin with 6–7 soft rays. The caudal fin is rounded. The anal fin has 9–10 soft rays.

The pygmy killifish is attractively colored. The back and sides are greenish brown and the belly whitish. There is a pronounced difference in the coloring of males and females. Males have 6–9 brownish vertical bars on the sides between the anal and caudal fins. A distinct black spot (ocellus) is found at the base of the caudal fin. Generally the sides are washed with blue-green and the fins vary from yellowish to intense yellow-orange. Females also have the caudal ocellus, but it is often surrounded by a brown ring. A prominent brown stripe extends from the caudal ocellus to the snout. There is usually a second ocellus, above the anus, in females. Juvenile fish of both sexes have the two brown ocelli and a dark band connecting them until the adult markings are acquired.

68

Similar Species.   The distinct caudal ocellus serves to identify this species.

Distribution.   The pygmy killifish occurs along the lower edge of the Atlantic Coastal Plain from the Ogeechee River in Georgia southward and westward through the panhandle of Florida to possibly the Escatawpa River drainage in Alabama. In the central part of peninsular Florida it ranges as far south as Tampa.

Habitat.   The pygmy killifish occurs in habitats with heavy plant growth, particularly in stream and lake margins or aquatic prairies. It is often seen in small schools in open water areas among vegetation. This species is one of the most common found in the Okefenokee Swamp, where it occurs in most aquatic habitats.

Biology.   The pygmy killifish is carnivorous, feeding mainly on small crustaceans and insects associated with vegetation on the bottom. Food items found in the stomach from the swamp and elsewhere support the observation made by others (1) that this species is not a surface feeder as are most related species of the genus *Fundulus* or *Gambusia*.
  Pygmy killifish have a complex courtship behavior after which they spawn in dense vegetation. Spawning probably occurs at least from early spring to late summer as breeding fish have been collected from April through August. Young spawned during late winter or early spring may be mature enough to breed in fall. In warmer waters reproduction may occur throughout the year. The pygmy killifish lives approximately 2 years.

Comments.   This is an attractive fish that does well in aquaria, where it makes a good subject for the study of spawning and courtship behavior. This killifish is the smallest oviparous (egg-laying) fish in the United States.

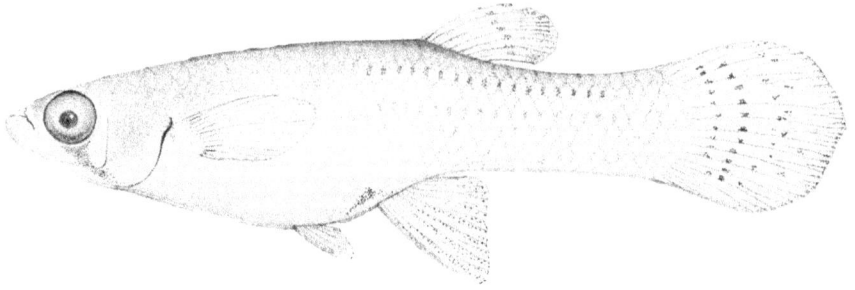

Female

Mosquitofish

Gambusia affinis (Baird and Girard)

Other Local Names.   Guppy, minnow, pieded minnow, pot gut, topminnow.

Scientific Name.   *Gambusia,* from the Latinized, provincial Cuban word for this species; *affinis,* from the Latin meaning "related to."

Description.   The mosquitofish is quite small. The females, which may attain almost 3 inches, are larger than the males, which seldom exceed 1 inch. The body is robust anteriorly, but it is abruptly reduced behind the dorsal and anal fins. The dorsal surface of the head is flattened. The mouth is terminal and markedly upturned. The caudal fin is rounded. The third ray of the anal fin is unbranched. A marked sexual dimorphism exists other than relative size differences. In females the origin of the dorsal fin is over the posteriormost anal fin ray; in males the origin of the dorsal fin is behind the posteriormost anal fin ray. In mature males the first few anal fin rays are greatly modified, forming a copulatory organ, the gonopodium. The abdomen of gravid females may be greatly distended.

The back and sides are olive- to yellow-brown. The scales are usually boldly outlined with dark pigment. Frequently there is a dark blotch beneath the eye. Also, in females, a dark "pregnancy" spot may be found on the side above the vent. There are usually 2–3 transverse rows of faint spots on the dorsal and caudal fins.

Similar Species. The mosquitofish can be distinguished from the closely related least killifish, which has a distinct, dark longitudinal band on the side and a series of dark vertical bars. Superficially, the mosquitofish resembles members of the topminnow family—including the pygmy killifish. In these, however, the origin of the dorsal fin is in front of or only slightly behind the origin of the anal fin.

Distribution. The species ranges naturally throughout most drainages of the southeastern United States from Delaware to Florida and west along the Gulf Coast to northeastern Mexico and north throughout much of the lower Mississippi River Valley. It has been widely introduced elsewhere, particularly in the Southwest. It occurs in both the St. Marys and Suwannee rivers.

Habitat. The mosquitofish is most abundant in shallow, quiet waters with heavy vegetation. It is very common in and around the Okefenokee Swamp, including lakes, ponds, prairies, canals, boatruns, and the quiet backwaters of sluggish streams.

Biology. The mosquitofish is largely restricted to the surface of shallow bodies of water, where it feeds on small crustaceans, aquatic insects and their larvae, and algae (5).

The species is viviparous, giving birth to live young rather than laying eggs. The breeding season extends throughout late spring and summer. Males court females almost continuously, and a female may produce 3–4 broods per year (47). Fertilization is internal and is accomplished by the transmission of sperm along a groove in the gonopodium, which is inserted into the female. The eggs develop inside the female and hatch in 2–3 weeks. As many as several hundred live young may be shed. These young grow rapidly and reach sexual maturity later in their first year. Females grow more rapidly, reach sexual maturity earlier, and live longer than males. Few mosquitofish survive their first year.

Comments. This species has been widely introduced for mosquito control because of its heavy predation on the aquatic larvae (47).

Gambusia is adept at invading new territory. It is not uncommon to find them in swollen drainage ditches or even in ruts in roads after or during a heavy rain.

There are two recognized subspecies—*G. a. holbrooki,* which occurs along the southeastern Atlantic Coast and peninsular Florida, and *G. a. affinis,* which occurs throughout the Mississippi River Valley and remaining Gulf Coast (51).

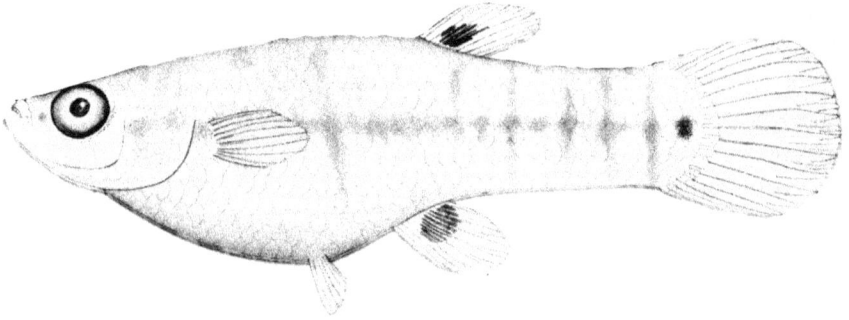

Female

Least Killifish

Heterandria formosa Agassiz

Other Local Names.   Minnow, topminnow.

Scientific Name.   *Heterandria*, from two Greek words for "different male"; *formosa*, from the Latin for "comely" or "pretty."

Description.   The least killifish is the smallest fish in North America, with females rarely exceeding 1 inch and males ¾ inch. The top of the head is flat. The snout is short and the terminal mouth is markedly upturned. The short body is robust anteriorly but abruptly reduced posterior to the dorsal and anal fins. The dorsal fin is short, and its origin is above (in females) or behind (in males) the middle of the anal fin. The caudal fin is rounded. As in the mosquitofish, the third ray of the anal fin is unbranched. In males the first few rays of the anal fin are markedly extended to form a gonopodium.

The back and sides are olive to brown. The belly is whitish. A distinct dark bar extends along the midside from behind the eye to the base of the caudal fin. Perpendicular to this are 6–9 dark vertical streaks. Pronounced dark spots are present on the bases of the dorsal, caudal, and anal fins of females. Males are similarly marked but lack the anal fin spot.

72

Similar Species. The least killifish can be distinguished from the closely related mosquitofish, which lacks the dark, longitudinal band and vertical bars. Superficially, the least killifish resembles the topminnows and pygmy killifish, but these lack the distinctly terminal and upturned mouth.

Distribution. This species is restricted to the margins of the Atlantic and Gulf coastal plains from South Carolina west to Louisiana. Other than in peninsular Florida it rarely ranges inland more than 100 miles. It is known from both the St. Marys and Suwannee rivers.

Habitat. The least killifish is found in many of the same habitats occupied by the mosquitofish and the cyprinodontids. Throughout its range it occurs in quiet waters with very heavy plant growth—including various drainages, swamps, ponds, the margins of large lakes, and quiet shallows of smaller tributary streams. This species has not been recorded in the Okefenokee National Wildlife Refuge but does occur in tributary creeks of the Suwannee River above Fargo. Although it is considered a freshwater fish, this species is commonly encountered in brackish waters throughout its range.

Biology. Like the mosquitofish, this species is largely restricted to the water surface, where it feeds on the larvae of aquatic insects, small crustaceans, and to a lesser extent on algae.

Breeding occurs from early spring through late summer (55). The species is viviparous, giving birth to live young. It differs from the mosquitofish, however, in that fewer numbers of young are shed (usually only 1–8) sequentially over several days (55). This results from a condition called *superfetation* in which a progressive series of eggs may reach maturity and are fertilized as more advanced embryos are shed. As a result, as many as six stages of development may occur in a single female at one time. The young develop rapidly and usually reach sexual maturity in their first year. Few survive the first year.

Comments. This fish is reported to be the smallest viviparous vertebrate known. Possibly for this reason it is a popular aquarium fish.

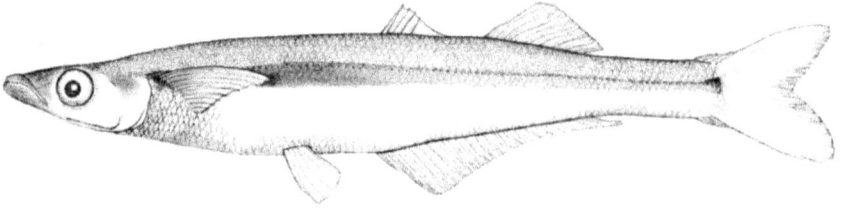

Brook Silverside

Labidesthes sicculus (Cope)

Other Local Names. Glassfish, needlenose, skip-jack, stick-minnow.

Scientific Name. *Labidesthes,* from two Greek words for "tongs" or "forceps" and "to eat," referring to the elongate jaws; *sicculus,* from the Latin meaning "dried," because the species is sometimes found in half-dried ponds.

Description. The brook silverside is an elongate and very slender small fish averaging 3–4 inches with a maximum length of 4½ inches. The head is long with a beaklike snout. The fish has 2 separate dorsal fins; the front one is short with 4–6 weak spines, and the second is longer with 1 spine and 10–11 branched rays. The caudal fin is deeply forked. The pectoral fins are also long, reaching almost to the pelvic fins.

  The back and sides are a pale greenish yellow. A bright silvery band extends along the midlateral sides. The belly is silvery white. The body is translucent, and the swim bladder and vertebral column can often be seen through the body musculature. The scales on the back may be slightly outlined in black.

Similar Species. This species is rarely confused with any other fish in the Okefenokee Swamp.

Distribution. This fish occurs from the southern Great Lakes throughout the Mississippi River drainage. It ranges along the Gulf Coast from Texas to Florida and north along the Atlantic Slope to South Carolina. It occurs in both the St. Marys and Suwannee rivers.

Habitat. The brook silverside is a surface dweller, rarely descending to more than a few feet in the water. The species has a strong tendency to school. The fish live in the quiet waters of lakes, prairies, boat runs, and the smaller tributary creeks. Habitat preference varies with the age of the fish. Hatchlings and young are usually found in deeper open water away from the water's edge (and away from predators such as wading birds). Adult fish prefer the water's edge. At night the large schools tend to break up, with individual fish being widely dispersed and resting on or near the surface. This species may be very abundant in appropriate habitat.

Biology. This species is most active during the day and on moonlit nights. The primary sources of food are microcrustaceans, but small aquatic insects and their larvae are also eaten (15, 46). The brook silverside very rarely feeds on other smaller fishes. Individual fish may leap out of the water to capture flying insects.

The spawning season in the Southeast apparently extends from March to September (55). The fish spawn in large schools usually in and around aquatic vegetation. Several males may actively pursue a gravid female, but usually only a single male will spawn with her. A characteristic courtship ritual takes place in which a male and female swim in alignment, the male above the female. The eggs are orange in color and, while not adhesive themselves, have a long adhesive filament, which permits the eggs to attach to aquatic vegetation or submerged debris. The eggs hatch in 7–10 days. Growth is extremely rapid, with the young attaining maximum size their first year (63). They spawn in their second year, but most do not survive more than 2 years.

Comments. The brook silverside has little importance to man, although larger specimens are occasionally used as bait for bass—however, they tend to die quickly.

This is a rather nervous little fish. The slightest disturbance will send individual fish and whole schools scurrying off, many of them leaping out of the water in their escape. They are more easily approached at night with the aid of a flashlight along the water's edge. They can even be caught by hand if approached from the front (55).

Although not yet verified by detailed study, it has been suggested there are two subspecies—*L. s. sicculus,* ranging throughout the mid-Mississippi River Valley northward, and *L. s. vanhyningi,* ranging in the lower Mississippi River Valley and along the Gulf and southeastern Atlantic coast (51).

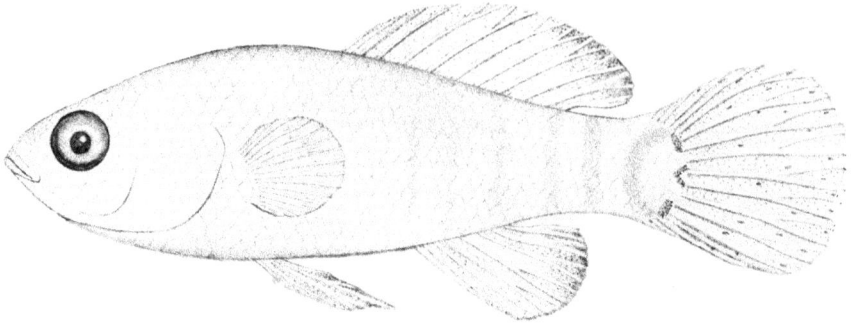

Everglades Pygmy Sunfish

Elassoma evergladei Jordan

Other Local Names. Pygmy sunfish.

Scientific Name. *Elassoma,* from the Greek for "smaller" or "less," a diminution; *evergladei,* the Latinized form of the swampy region of Florida in which it was first found.

Description. This diminutive sunfish earns its name, pygmy, because of its small size, usually ¾–1½ inches in length. The body is shaped like that of a sunfish or bream—oblong and laterally compressed, only much smaller. The top of the head is evenly scaled. The mouth is small, reaching only to the anterior margin of the eye. The dorsal fin is low, with 3–4 spinous and 8–9 soft rays. The body is deepest at the level of the anteriormost part of the dorsal fin and decreases toward the rounded caudal fin. The anal fin has 3 spines and 4–5 soft rays. A lateral line is absent.

The coloration of this fish varies but is usually dark olive brown with darker blotches that appear as faint vertical streaks. The dorsal and anal fins have several irregular rows of dark pigment; sometimes there are reddish spots at their bases. The caudal fin usually has two dark spots at its base. Breeding males may have an irregular mottling of bright iridescent blue.

Similar Species. This species superficially resembles the Okefenokee pygmy sunfish; the latter, however, lacks scales on the top of the head and has 10–13 soft rays in its dorsal fin and 6–8 soft rays in its anal fin. The Everglades pygmy sunfish also resembles the young of a number of centrarchid (bream) species, but these all have more (between 6 and 13) dorsal spines. Another species of pygmy sunfish, the banded pygmy sunfish, is found well to the south of the swamp watershed in both the St. Marys and Suwannee rivers. This species is characterized, however, by 10–12 distinct vertical bars along the sides and a distinct dark round spot under the anterior part of the dorsal fin.

Distribution.   The species occurs along the Atlantic Slope from North Carolina south throughout Florida and west to Mobile Bay, Alabama. It lives throughout the St. Marys and Suwannee river drainages.

Habitat.   The species is an inhabitant of all types of quiet waters with dense vegetation. It is widely distributed throughout the Okefenokee Swamp in lakes, prairies, boat runs, small tributary creeks, as well as sloughs of larger streams and rivers. After especially heavy rains and flooding, the fish may even be found in deep ruts and ditches of dirt roads quite some distance from permanent water. It is sometimes encountered in sphagnum bogs.

Biology.   The diet of this pygmy sunfish includes primarily small crustaceans, insect larvae, and nymphs.

In Florida, breeding may occur throughout the year, although it may be concentrated in late winter through early summer. Spawning may occur several times over a period of a few weeks. A small number of eggs, usually less than 100, are shed by the female during each spawning. This species does not build a nest; rather, the eggs are deposited in and around submerged vegetation and other debris to which they adhere. After being fertilized by the male, eggs hatch in approximately 1 week. The young attain sexual maturity their first year. Maximum life span is usually 3 years or less.

Comments.   This species is very popular among aquarists because of its small size and generally docile nature. This fish is abundant in appropriate habitat.

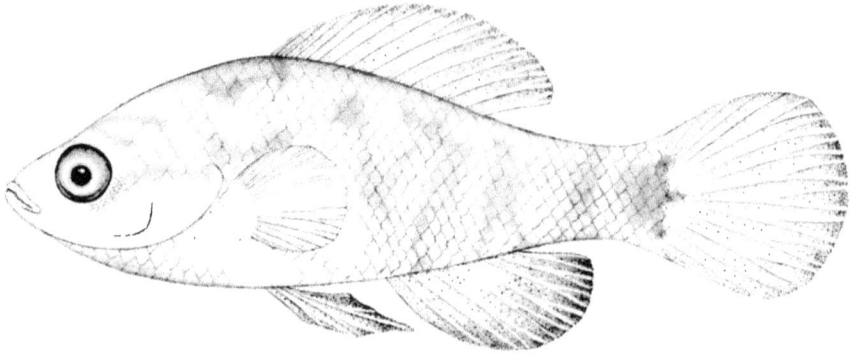

Okefenokee Pygmy Sunfish

Elassoma okefenokee Böhlke

Other Local Names.   Pygmy sunfish.

Scientific Name.   *Elassoma,* from the Greek for "smaller" or "less," a diminution; *okefenokee,* named after the Okefenokee Swamp from which it was first described.

Description.   This small fish ranges from ¾–1½ inches. Its body shape resembles that of a small sunfish or bream—oblong and laterally compressed. The top of the head lacks scales. The mouth is small. The dorsal fin is long and low, with 3–5 spinous and 10–13 soft rays. As in other pygmy sunfish, the body is deepest at the level of the anteriormost part of the dorsal fin and decreases toward the rounded caudal fin. There are 3–5 anal spines and 6–8 anal soft rays. A lateral line is absent.

The back and sides of the body are light brownish and have a pattern of darker brownish bars and blotches. Two longitudinal rows of small white spots are found on the basal portion of the caudal fin. Males, particularly when breeding, have scattered, irregularly shaped, iridescent blue blotches. The dorsal, caudal, and anal fins of males may also have blue crescents.

Similar Species.   The Okefenokee pygmy sunfish can be confused with the Everglades pygmy sunfish and juvenile fish of a number of species of bream in the centrarchid family. The former, however, has scales on the top of the head and only 8–9 soft rays in the dorsal fin. All members of the centrarchid family have between 6 and 13 dorsal spines and a lateral line. Another pygmy sunfish—the banded pygmy sunfish—occurs south of the Okefenokee Swamp watershed, but it is very distinct (see similar species in preceeding account).

Distribution. This species has a very limited distribution ranging from the Altamaha River in Georgia west to the Choctawhatchee River in Florida. It occurs in both the St. Marys and Suwannee rivers.

Habitat. Like other pygmy sunfish, this species occurs only in quiet waters with dense plant growth. It usually lives in lakes, prairies, canals, sloughs, creeks, and backwater pools of larger streams and rivers. Its habitat preferences appear to be somewhat similar to those of the Everglades pygmy sunfish. In many areas of the Okefenokee Swamp these two species occur in the same areas and are often taken together in nets. The species is most commonly found in areas of overhanging plants and rootlets in relatively deep water.

Biology. The species feeds on small crustaceans, as well as on larval aquatic insects and nymphs.

Specific information on the breeding biology of the Okefenokee pygmy sunfish is not available. Presumably, however, its habits are very similar to that of the Everglades pygmy sunfish.

Comments. Unfortunately, very little is known about the biology of the fish named after the Okefenokee Swamp because it was only recently (1956) discovered (10). The species was first recognized as a distinct species not by a professional ichthyologist but rather by an avid aquarist, L. Short, who had taken specimens from Kettle Creek, near Waycross, Georgia, as well as from other sites in the Okefenokee Swamp (82).

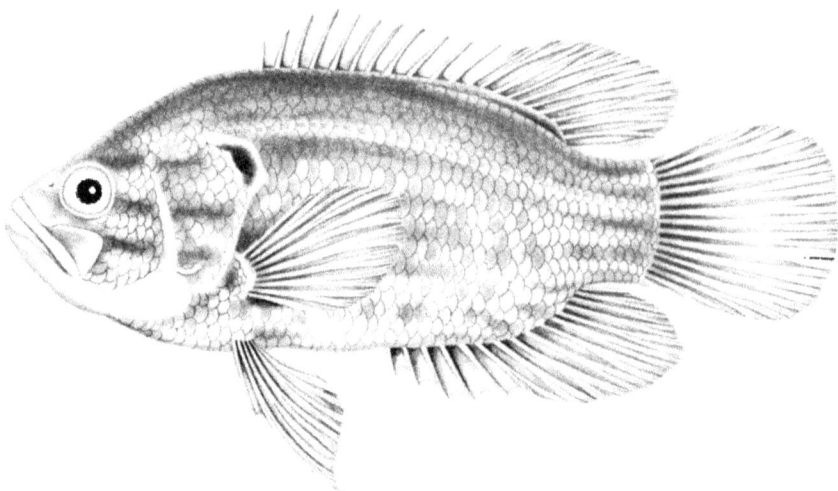

Mud Sunfish

Acantharchus pomotis (Baird)

Other Local Names. Bream, mud bass, mud perch.

Scientific Name. *Acantharchus,* from two Greek words referring to the "spiny vent"; *pomotis,* named from a genus of sunfish synonymous with *Lepomis.*

Description. The mud sunfish is a stout-bodied but very noticeably oblong sunfish that reaches a maximum length of 7–8 inches. The snout is short. The broad maxilla extends almost to the posterior part of the eye. The dorsal fin is not divided by a deep notch and has 11–12 spinous and 10–11 soft rays. The caudal fin is distinctly rounded, contributing to the oblong shape of the fish. There are 5 spinous and 10 soft anal rays.

Coloration is a rather uniform grayish green, and no bright colors are present. There are 4–6 dark horizontal stripes—which are very distinctive, even in juveniles—along the sides. Several dark horizontal bars are present on the head, the lowest of which extends onto the mandible. There is a dark opercular spot.

Similar Species. The combination of oblong shape, rounded caudal fin, and distinct horizontal bars allows this species to be easily distinguished from other sunfish.

Distribution. The mud sunfish occurs along the Atlantic Coastal Plain from New York southward through the Florida panhandle to the St. Marks River. In Georgia it is widely but sparsely distributed throughout the lower coastal plain from the Savannah through the Ochlockonee river drainages.

Habitat. The mud sunfish is found in habitats characterized by silt, mud, and abundant vegetation. In the Okefenokee Swamp, it has been sporadically collected near dense aquatic vegetation.

Biology. This species is very secretive and its feeding preferences are poorly known. It presumably feeds on aquatic insects, their larvae, and crustaceans. Adult fish frequently rest head down in weeds and are most active at night.

Spawning occurs from early fall to late winter in the Okefenokee Swamp region. Typical circular sunfish nests have been observed in sand and mud surrounded by aquatic plants. The fish may live as long as 7 years (57).

Comments. The mud sunfish is not generally considered a game fish, although adults do grow large enough to be used as a panfish. This species is so secretive that some local fishermen who had fished throughout their lives in a pond in which the mud sunfish lived did not recognize the fish or even have a common name for it.

A distinctive subspecies of mud sunfish, *A. p. mizelli*, was described in 1945 from the Suwannee River drainage. This subspecies has not generally been accepted in subsequent literature. Specimens from the Gulf drainage rivers are reported to differ somewhat from those of the Atlantic drainages (51). In the Okefenokee Swamp this lack of acceptance of the subspecies may seem unfortunate because H. W. Fowler, who described the subspecies, named it after Hamp Mizell, one of the "swampers" made famous by Francis Harper's popular book, *Okefinokee Album* (35). Mizell Prairie is also named after the family.

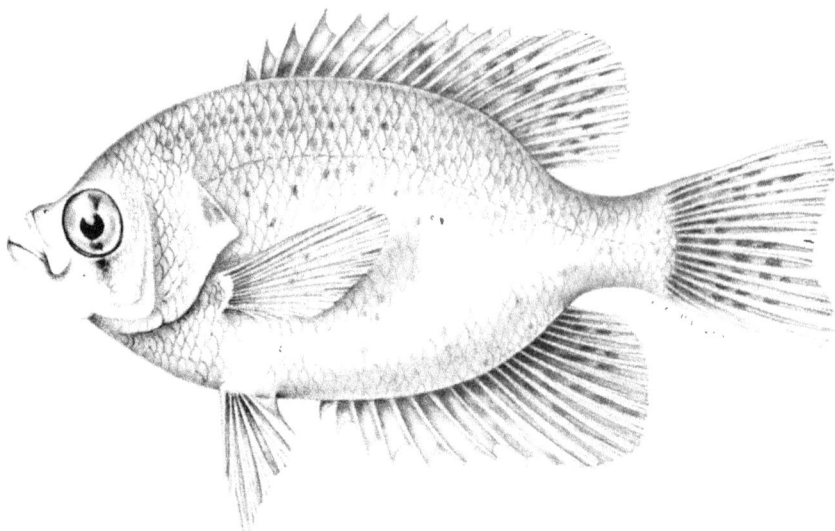

Flier

Centrarchus macropterus (Lacepède)

Other Local Names.   Bream, brush bream, goggle-eye, sandflirter, sandperch, shiner, thin-gizzard.

Scientific Name.   *Centrarchus,* from two Greek words for "anal spines"; *macropterus,* from two Greek words for "long fin."

Description.   The flier is a very deep-bodied, laterally compressed sunfish that is almost circular in appearance. It is a small fish, ranging between 3 and 5 inches in length. Rarely a specimen may attain 7 inches. The head is small and the snout short. In lateral view the head is markedly concave. The mouth is moderately large, the upper jaw extending to the middle of the eye. The free margin of the preopercle is strongly serrate. There are 11–13 spinous and 12–14 soft dorsal rays with no deep notch between them. The caudal fin is slightly forked. The anal fin is long with 7–8 spinous and 13–15 soft rays.

The back and sides of this very attractive sunfish are yellowish green to olive and the belly yellowish. Below the lateral line is a series of dark brown spots forming interrupted longitudinal lines. Usually a dusky, almost teardrop-shaped, vertical bar may be found beneath the eye. Young individuals have a dark spot surrounded by a reddish orange halo on the posterior soft dorsal rays, but this is usually absent in adult fish. Both the soft dorsal and anal rays are slightly mottled with narrow dark bars.

Similar Species.   The flier superficially resembles other sunfish, particularly the black crappie; however, it is easily distinguished from the black crappie because the flier has 11–13 spinous dorsal rays.

Distribution.   The flier ranges along the Atlantic Slope and Gulf coasts from Maryland to Texas and up the Mississippi River Valley to southern Illinois. It is fairly common in the local region, occurring widely in both the St. Marys and Suwannee river drainages.

Habitat.   This is the most common sunfish in the Okefenokee Swamp. Typically it prefers relatively sluggish waters with abundant vegetation. In the Okefenokee Swamp, it is known from lakes, prairies, canals, and boat runs, as well as from backwater pools and sloughs of tributary streams and main rivers. As with most other sunfish, individual specimens of this species tend to remain in and around dense vegetation and submerged debris, especially during the day. During summer months, the species has no pronounced tendency to school, though schooling may be seen in winter months.

Biology.   Small crustaceans, aquatic insects, and filamentous algae make up the greatest part of the flier's diet. Larger fliers may also feed heavily on the young of other fishes when the latter are abundant (20).

The specific breeding habits of the flier are not known in detail. Their habits are presumed to be similar to those of other sunfishes. The breeding season probably extends from February to as late as October. Fliers are colonial nesters, and males are known to guard the eggs and fry. The young mature at approximately one year of age (2½–3 inches). Thereafter they grow more than an inch per year (20). Longevity is 6–7 years. Females tend to be larger than males and generally live longer as well.

Comments.   As the most common sunfish in the Okefenokee Swamp, the flier makes up a considerable portion of sunfish creel records. Despite the comparatively small size, the species is a very popular sport fish. They readily take artificial and live bait. In fact, the rapidity of their strike is the reason for their common name, flier.

Blackbanded Sunfish

Enneacanthus chaetodon (Baird)

Other Local Names.   Banded sunfish, bream.

Scientific Name.   *Enneacanthus,* from two Greek words for "nine" and "spine"; *chaetodon,* named for a genus of marine fish with similar crossbands.

Description.   The blackbanded sunfish is a small sunfish, growing to a maximum size of only 2½–3 inches. It is very deep-bodied and highly compressed. The mouth and head are small. The eyes are large. The emarginate dorsal fin has 9–10 (usually 10) spinous and 12 soft rays with no deep notch between them. The caudal fin is convex. The anal fin has 3 spinous and 12 soft rays.

   The coloration of this fish is very striking. The ground color is dirty white or pale straw to brassy yellow. There are 6–8 prominent, black vertical bands, the first of which passes through the eye. The third vertical band extends through the membrane of the first three spinous rays of the dorsal fin and ventrally onto the pectoral fin. The sides are also variously mottled with darkened scales. In life, the leading edge of the pelvic and anal fins are margined with orange. The opercular spot is black but may appear to be two spots due to a pale crescent-shaped center. Differences between the sexes are not apparent.

Similar Species.   Both the banded sunfish and the bluespotted sunfish might be confused with the blackbanded sunfish. The distinctive combination of typical brassy yellow color and prominent black bands passing through both the eye and the first three membranes of the spinous dorsal fin, however, will permit instant identification of adult blackbanded sunfish. Juvenile fish of all three species present special problems for identification, but in general even juvenile blackbanded sunfish are distinctively colored and marked as opposed to those of the other two species.

Distribution.   The blackbanded sunfish occurs from the Pine Barrens region in coastal New Jersey southward along the Atlantic Coastal Plain into northern Florida. It has a spotty distribution throughout the coastal plain of Georgia and northern Florida. It occurs throughout the central portion of the Okefenokee Swamp. Except for populations in North Carolina and New Jersey coastal plains, this fish is uncommon to rare throughout its range and densities are often low.

Habitat.   The blackbanded sunfish prefers quiet waters between 2 and 3 feet deep with abundant plant growth. In the Okefenokee Swamp it has been taken in only a few localities scattered across the central portion of the swamp.

Biology.   This species tends to aggregate in waters with heavy vegetation. The blackbanded sunfish is considered to be primarily a nocturnal bottom feeder (79). The diet consists primarily of insects and crustaceans, particularly the larvae of chironomids, caddisflies, and dragonflies.

Very little information on breeding is available, though spawning is thought to occur in late winter to early spring. Males apparently construct a small nest in which the eggs are deposited (13). The male guards the nest and the eggs, which hatch in 2 days—after which there may be subsequent spawnings (69). Blackbanded sunfish live approximately 4 years.

Comments.   Populations of this species in Georgia and Florida were formerly recognized as a separate subspecies, *Enneacanthus c. elizabethae*. More recent studies, however, have indicated subspecific distinction is not warranted (85).

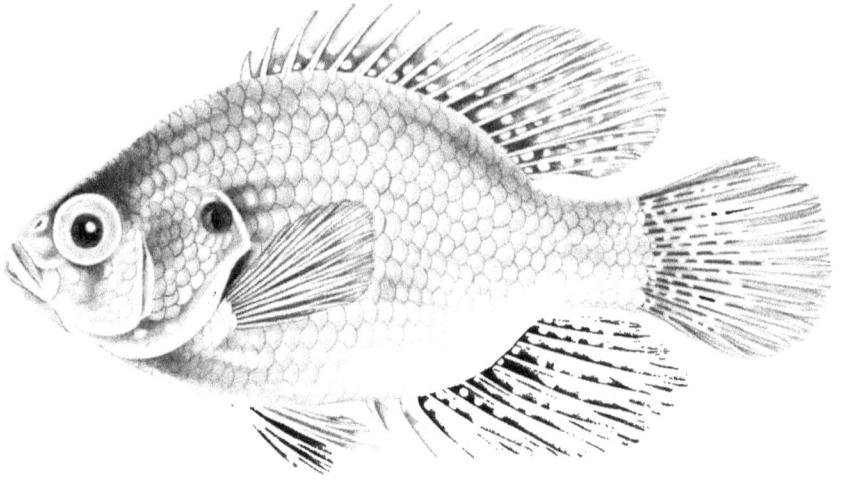

Bluespotted Sunfish

Enneacanthus gloriosus (Holbrook)

Other Local Names. Bream, speckled perch.

Scientific Name. *Enneacanthus,* from two Greek words for "nine" and "spine"; *gloriosus,* from the Latin for "glorious."

Description. The bluespotted sunfish is small and rarely exceeds 2–2½ inches throughout its range, although most specimens in the Okefenokee Swamp region are smaller. It is less deep-bodied and more elongate than other members of the genus. The mouth and head are small. There are 9–10 spinous and 10–11 soft dorsal rays with no deep notch between them. The dorsal fin is not emarginate. The caudal fin is convex. The anal fin usually has 3 spinous and 9–10 soft rays.

The bluespotted sunfish is brightly colored. Its ground color is dark olive green with bright blue spots arranged in irregular horizontal rows along the sides extending through the fins and even onto the head. These spots are more apparent in males. In females the spots are larger but duller and less distinct than in males. An oblique bar is present below the eye. The opercular spot is small and is pearly blue in color with a darker blue margin. Vertical crossbars are present in many specimens, but may become obscured in larger specimens. In juvenile fish the vertical crossbars are not complete, but are broken up, giving a jagged appearance. Males are typically more brightly colored than females.

Similar Species.   All species of *Enneacanthus* might be confused with each other. In general, the blackbanded sunfish is distinctly marked. The banded sunfish has distinct vertical bars, lacks blue spots, has a larger opercular spot, lacks the oblique bar beneath the eye, and is more ovoid in appearance.

Distribution.   This species is common along the Atlantic Slope from the Hudson River south throughout Florida and west along the Gulf Coast to Mobile Bay. It is fairly common in the Georgia Coastal Plain and is found throughout the Okefenokee Swamp.

Habitat.   The bluespotted sunfish is found over muddy or sandy bottoms in the marginal quiet waters with thick vegetation of backwater pools, roadside ditches, or aquatic prairies. In general this species does not require the critical combination of water depth and vegetation needed by the blackbanded sunfish. Also, it is fairly tolerant of a range of acidity, occurring in acidic blackwater as well as unstained alkaline streams. It is less ecologically specialized than the other two *Enneacanthus* species.

Biology.   This species tends to aggregate in large numbers in the dense vegetation near the water's edge. They feed predominantly on small crustaceans and aquatic insect larvae and pupae.

The breeding season extends throughout spring and summer from April to October in the south (55). Conventional sunfish nests may be constructed or the adhesive eggs may be deposited on or in masses of plants typically in fairly shallow water. The number of eggs released ranges from 84 to 635 (55).

Comments.   This species is distinct from the closely related banded sunfish, *E. obesus.* However, the two species occur together throughout much of their range, and there is reported to be a rather confusing intermediacy in some populations indicating the possibility of hybridization (51). Palmer and Wright note some difficulty in distinguishing the species in their original studies of the fishes of the region (64). We have not observed any evidence of hybridization in our studies.

Banded Sunfish

Enneacanthus obesus (Girard)

Other Local Names.   Bream, little sunfish.

Scientific Name.   *Enneacanthus*, from two Greek words for "nine" and "spine"; *obesus*, from the Latin for "fat."

Description.   The banded sunfish, like the other species of *Enneacanthus*, is a small fish seldom exceeding 2½ inches in length. This species is deep-bodied and compressed, but also presents an ovoid appearance. It generally has 9 dorsal spinous and 10 dorsal soft rays, with no deep notch between the spinous and soft portions. The anal fin usually has 3 spinous and 10 soft rays. The lateral line is usually incomplete, ending under the soft part of the dorsal fin.

The banded sunfish is distinctively marked. The sides are crossed by 5–8 sharply demarcated vertical crossbars, which are convex toward the anterior. The black opercular spot is larger than the diameter of the pupil. A vertical bar is present below the eye, but it is not as uniformly wide as that of the bluespotted sunfish. The background color is olivaceous. Males are usually more distinctly marked than females. Females are typically smaller in size than males (55).

Similar Species.   The banded sunfish might be confused with both the black-banded and the bluespotted sunfish. The brassy yellow color, and the prominent, straight vertical bars on the sides of the blackbanded sunfish serve to distinguish it from the banded sunfish, which has curved bars and an olivaceous color. The bluespotted sunfish lacks well-defined vertical bars, even in juvenile specimens, and is covered with irregularly arranged rows of blue spots.

Distribution.   The banded sunfish occurs from southern New Hampshire southward along the Atlantic Coastal Plain into southern Georgia and northern peninsular Florida and westward through panhandle Florida to the Perdido River drainage.

Habitat.   This species is often found in large groups in waters with heavy vegetation and little or no current, and usually with sandy or silty bottoms. In the Okefenokee Swamp it lives primarily among the abundant plant growth of prairies and margins of boat trails. It occurs often in drainage ditches connected with pine flatwoods streams. It remains in these areas even at times when water level is very shallow (12 inches or less). Frequently the banded sunfish occurs with the bluespotted sunfish. Both fish are fairly tolerant of nutrient-poor or stagnant waters. The banded sunfish is reported to be somewhat more restricted in its distribution than the bluespotted sunfish in some areas of the Southeast (55), though it occurs more widely in the Okefenokee Swamp than does either of the other species of *Enneacanthus*.

Biology.   The banded sunfish feeds primarily on aquatic insects, crustaceans, and molluscs.
Spawning has not been observed in the species, but in Florida they apparently breed from March through September (55). Juvenile fish have been collected in the Okefenokee Swamp from April through October, suggesting a similar extended breeding season here. The species probably breeds in waters with abundant plant growth.

Comments.   This species and the bluespotted sunfish occur together in many areas of their ranges, and apparently individual specimens in some of these populations are intermediate in appearance between the two species (51). We have not observed this hybridization in any specimens from the Okefenokee Swamp.

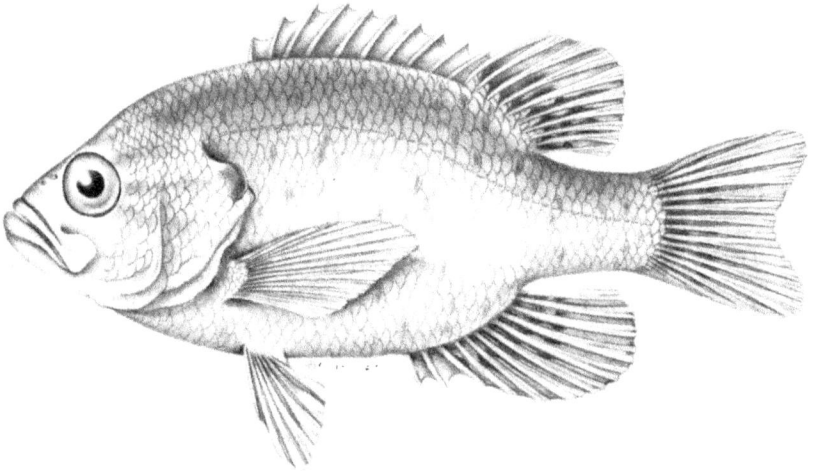

Warmouth

Lepomis gulosus (Cuvier)

Other Local Names.   Goggle-eye, more-mouth bream, red-eye stumpknocker.

Scientific Name.   *Lepomis*, from two Greek words referring to the scaled gill cover; *gulosus*, from the Latin for "large-mouthed."

Description.   The warmouth is a shallow-bodied, robust sunfish. Most are 5–7 inches in length, though some grow to 10–11 inches. The mouth is large and the posterior margin of the upper jaw extends beyond the middle of the eye. The tongue has a small patch of teeth. The dorsal fin is elongate and has 9–10 spinous and 9–10 soft rays with no deep notch between them. The caudal fin is slightly forked. The anal fin has 3 spinous and 8–9 soft rays.

   The back and sides are a dark greenish yellow to a brassy olive brown mottled with darker brownish blotches. These blotches are prominent in breeding males but may be absent in fishes from very turbid waters. The belly is yellowish to brassy. From 3 to 5 dark bars radiate backward from the snout and eye (the iris of which is red, hence its common reference as "red-eye"). The posterior margin of the opercle bears a prominent dark spot. The dorsal, caudal, and anal fins are strongly banded. Breeding males may have a bright orange or reddish spot at the base of the posteriormost dorsal rays.

Similar Species.   The warmouth can easily be distinguished from other members of the sunfish family because it has: three anal spines, an emarginate caudal fin, no deep notch between the spinous and soft portions of the dorsal fin, and 3–5 dark bars radiating backward from the eye and snout.

Distribution.   The species is widely distributed throughout most of the eastern United States south of the Great Lakes west to the Mississippi Valley. On the Atlantic Slope it occurs from Chesapeake Bay south throughout Florida and west along the Gulf Coast through Texas. It has been widely introduced west of the Rocky Mountains.

Habitat.   The warmouth prefers quiet, dark waters in areas with muddy or silty bottoms. It is most common in and around aquatic vegetation or other submerged debris such as roots and stumps, which it uses for cover. It is sufficiently common around submerged stumps to have earned the title "stumpknocker," although this name is generally applied to the spotted and red-ear sunfish. In the Okefenokee Swamp the warmouth occurs in a variety of habitats, including lakes, prairies, boat runs, and canals, and less commonly in the backwater ponds and pools of tributary streams. In general it is more tolerant of muddy, silty water than are other sunfishes.

Biology.   The warmouth is generally considered more piscivorous than other sunfishes. In addition to fish it consumes crayfish as well as aquatic insects and larvae (34, 50). Feeding activity is concentrated in the early morning hours.

The breeding season extends from March to as late as September, although the warmouth spawns most frequently in early spring. Nests are prepared by the male in shallow, marginal areas that are choked with aquatic vegetation or other cover. As many as 50,000 eggs are deposited. These hatch usually in less than two days, and the fry, still guarded by the male, do not leave the nest for about a week. Growth is approximately 1½ inches per year. Sexual maturity usually is reached the second year. The warmouth lives 6–7 years.

Comments.   The relatively small size of the warmouth precludes its being a popular sport fish, although it is an excellent panfish. The warmouth readily takes a baited hook and flies as well as live bait, and the initial strike is quite vigorous. This species is known to hybridize with other members of the sunfish family, including bluegill, largemouth bass, and black crappie.

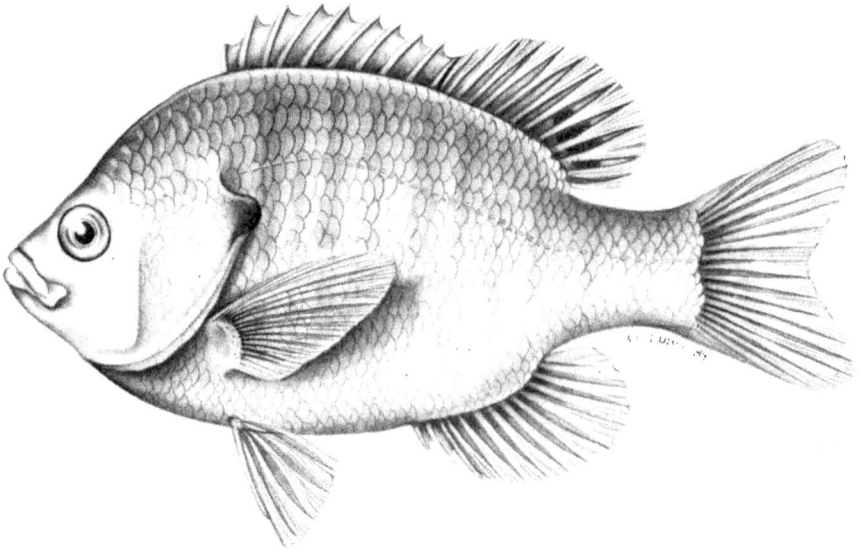

Bluegill

Lepomis macrochirus Rafinesque

Other Local Names. Bream, copper-belly, pond perch, sun perch.

Scientific Name. *Lepomis,* from two Greek words referring to the scaled gill cover; *macrochirus,* from two Greek words meaing "large hand," in reference to its body shape.

Description. This is a deep-bodied, slab-sided sunfish that reaches 7–9 inches. Large specimens may exceed 12–14 inches. The mouth is small, the posterior margin of the upper jaw never extending beyond the front of the eye. The flap of the operculum is thin and flexible. The long, somewhat pointy, pectoral fins usually extend beyond the eye if bent forward. There are 10 spinous and 10–11 soft dorsal rays with no deep notch between them. The caudal fin is weakly emarginate. The anal fin has 3 spinous and 10–12 soft rays.

In the Okefenokee Swamp the back and sides are typically olive green with a metallic greenish blue to brassy luster. There are usually 6–8 darker vertical bars along the sides. These are most prominent in juvenile fish and may be lacking in adult fish. The belly is whitish yellow to orange. The chin and lower part of the operculum are purplish blue. The opercular flap is black, and there is a dark blotch on the posterior base of the dorsal fin. Breeding males are colorful, with a bright orange to rusty red breast and a deep bluish sheen along the sides and back.

Similar Species. The bluegill can be distinguished from most other members of this family by: its three anal spines, the emarginate tail, no deep notch between the spinous and soft dorsal rays, and the flexible opercular flap. It is sometimes difficult to distinguish from the dollar sunfish, *L. marginatus.* The latter, however, lacks the vertical bars, has no dark spot on the dorsal fin, and has a distinct greenish white margin on the opercular flap.

Distribution. The bluegill's natural range includes most of the eastern United States. Today, as a result of extensive introductions, it occurs nearly throughout the nation and south into northern Mexico.

Habitat. The bluegill is a common but not abundant fish in the Okefenokee Swamp. It prefers the clean waters with heavy plant growth of ponds, lakes, and backwater pools and sloughs of rivers and tributary streams.

Biology. During the day, smaller bluegill may be found near shore vegetation while larger ones tend to remain in deeper, open water. In the morning and evening, however, the larger bluegill move toward the shoreline, where feeding activity is most concentrated. The diet of the bluegill is varied. The bluegill primarily feeds on aquatic insects, crustaceans, fish eggs, and smaller fishes but also consumes considerable amounts of plant material (55, 83).

In the Okefenokee Swamp region the bluegill spawns from May until October. Males construct small (diameter about twice the fish's length) nests in relatively shallow water 1–2 feet deep. More than 40,000 eggs may be laid by a single female. These eggs are guarded by the male, but only until they hatch (3–5 days). Growth is rapid—about 2 inches the first year and 1–1½ inches in each succeeding year. Bluegill live approximately 5 years.

Comments. The bluegill, a popular food and game fish, is possibly the most common "pond type" fish in the United States (27). Three subspecies of bluegill are recognized (42). *Lepomis macrochirus purpurescens* occurs throughout the Atlantic Slope from Virginia through Florida. The bluegill commonly hybridizes with other sunfishes, particularly with other members of the genus *Lepomis* (2).

Dollar Sunfish

Lepomis marginatus (Holbrook)

Other Local Names. Bream, long-eared sunfish.

Scientific Name. *Lepomis,* from two Greek words referring to the scaled gill cover; *marginatus,* from the Latin referring to the light-margined gill cover.

Description. The dollar sunfish is a small, but deep-bodied species. Adult fish range from 4 to 5 inches in length. The mouth is small, never extending beyond the front of the eye. The opercular flap is thin and flexible, and may be doubled over without breaking the opercular bone. The dorsal fin is continuous, without a deep notch between spinous and soft rayed fins. There are 10 spines and 11 soft rays in the dorsal fin. The caudal fin is weakly emarginate, and the anal fin has 3 spinous and 10–12 soft rays.

The dollar sunfish is an attractive fish. In adult specimens the opercular flap is margined with a greenish white border. The opercle and cheek are often marked with wavy, bluish green lines that may extend to the tip of the snout. The sides of adult fish may also be marked with greenish white spots on the scales. The pelvic fins may be milky in color, and the other fins are not clear but have the center portions of the rays suffused with olive green, fading to a lighter, creamy-colored belly. The sides are not banded and do not have dark spots.

Similar Species. The dollar sunfish is related to the longear sunfish, *Lepomis megalotis,* and resembles it closely (72)—however, the latter species does not occur in the coastal plain of Georgia. Small specimens of the warmouth, flier, bluegill, and spotted sunfish might be confused with the dollar sunfish. The dollar sunfish, however, lacks teeth on the tongue and bands or dark spots on the side.

Distribution. The dollar sunfish is found along Atlantic coastal drainages from North Carolina southward through peninsular and panhandle Florida. It occurs westward to Texas and north through the Mississippi embayment to Kentucky. In Georgia, it is widely distributed across the coastal plain below the Fall Line.

Habitat. The dollar sunfish occurs in waters with little or no current and often in those connected with swamps. In the Okefenokee Swamp region, the dollar sunfish has been collected in marginal pools along the St. Marys River and in some marginal areas of upland drainage streams. It has not been collected in the interior of the swamp.

Biology. Little is known about the biology of this species. The primary food source includes the larvae of aquatic insects and small crustaceans as well as molluscs (55).

Breeding fish have been reported from April to September from the St. Johns River in Florida (55). No additional information on reproduction is available.

Comments. The dollar sunfish is too small to be of importance as a game fish. Its small maximum size and bright coloration do make it an attractive aquarium fish.

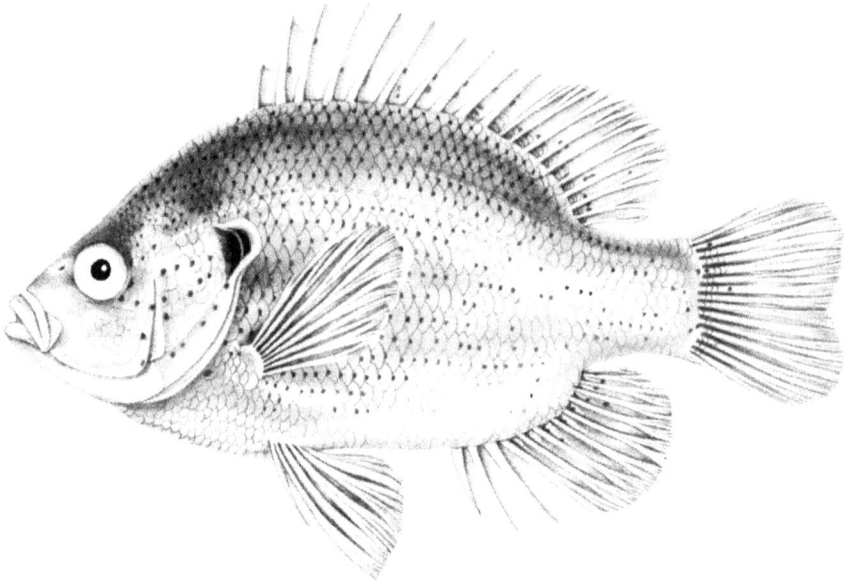

Spotted Sunfish

Lepomis punctatus (Valenciennes)

Other Local Names.   Bream, log perch, stumpknocker.

Scientific Names.   *Lepomis,* from two Greek words referring to the scaled gill cover; *punctatus,* from the Latin for "spotted."

Description.   The spotted sunfish is small, normally 2–6 inches in length, but grows to a maximum of 8 inches. It is a thick, deep-bodied fish with an oblong, nearly elliptical shape. The mouth is moderately large, with the upper jaw reaching just past the front of the eye. The operculum is stiff to the margin and not fimbriate. The dorsal fin has 9–11 (usually 10) spinous and 10–11 soft rays. The two portions are not deeply notched between. The caudal fin is slightly forked. The anal fin has 3 spinous and 10–12 soft rays. The pectoral fin is short and rounded and does not reach past the front of the eye when bent forward.

The back and sides are olivaceous, and the belly is yellowish to white. There are longitudinal rows of light spots along the sides. These spots are pale yellow-orange in females or reddish in males, but may be absent in juvenile fish. The cheek is dark green and the opercular flap is black with a light-colored margin. The fins are usually dusky-colored and lack distinctive markings. Breeding males, however, have distinctly black pelvic fins that they frequently extend while defending the nest.

Similar Species.   The spotted sunfish can be distinguished from other sunfish by: its three anal spines, an emarginate tail, no deep notch between the spinous and soft dorsal rays, and a stiff operculum.

Distribution.   The range extends along the Atlantic and Gulf coasts from North Carolina to Texas and up the lower Mississippi Valley.

Habitat.   The spotted sunfish is usually considered a stream species, but it occurs in sloughs, lakes, and swamps. In the Okefenokee Swamp it is known from larger streams, canals, and lakes. It is most common, if not abundant, in areas of dense vegetation and debris, particularly near submerged logs, fallen trees, or stumps (hence, its common name, stumpknocker).

Biology.   The bulk of the spotted sunfish's diet consists of insects and crustaceans, although plant matter, including considerable plant debris, is also eaten (55).

The breeding season in the Southeast extends from March to November (55). Solitary circular nests are constructed by the males, which prefer sand or gravel bottoms in shallow waters with heavy vegetation and submerged debris. Each male guards his nest and eggs vigorously. The eggs, which are reported to be blue in color, hatch in 2–3 days. Young reach sexual maturity the following year at approximately 2 inches in length. Life span is 4–5 years (16).

Comments.   Because of its small size, the spotted sunfish is not much sought after by fisherman. It is, however, frequently plentiful in small streams and is easily taken on light tackle using dry flies, worms, and cut bait. Despite their small size, they are, like most sunfish, an excellent panfish.

Two well-defined subspecies are recognized: *L. p. miniatus* throughout the Mississippi River Valley and *L. p. punctatus* along the Atlantic Slope and west along the Gulf Coast of Florida. The two subspecies share a zone of integradation in extreme western Florida and probably Alabama (51).

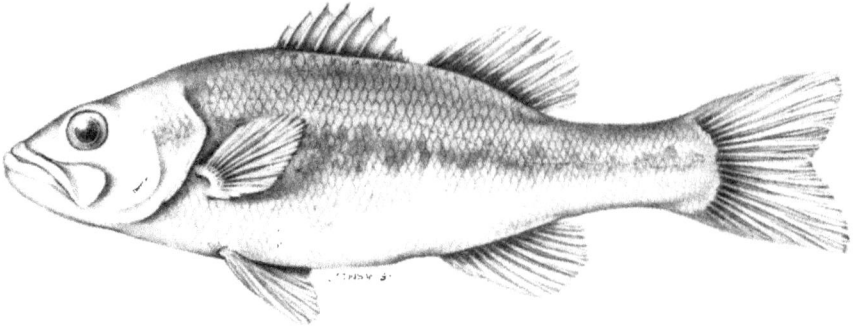

Largemouth Bass

Micropterus salmoides (Lacepède)

Other Local Names.   Bass, black bass, chub, line-side bass, trout, welshman.

Scientific Name.   *Micropterus,* from two Greek words for "small fin"; *salmoides,* from the Latin for "trout" and the Greek for "like," hence troutlike. Actually, the use of "small fin" hardly seems appropriate for these basses. The specimen from which Lacepède described the genus, however, had apparently suffered an injury to the posterior rays of the dorsal fin, making it appear short and detached.

Description.   The largemouth bass is a robust and elongate fish much less laterally compressed than other members of the family. It reaches a maximum length of approximately 28 inches, although most adults are 7–15 inches in length. The mouth, as indicated by the name, is large, with the upper jaw extending beyond the rear margin of the eye. The operculum is bony to the rear margin and does not extend to form a flap. There are 10–11 spinous and 12–13 soft dorsal rays. The spinous and soft dorsal rays are separated by a distinct, deep notch. The caudal fin is slightly forked. The anal fin has usually 3 spinous and 10–12 soft rays.

The upper portion of the body is dark green to olive. The lower sides and belly are whitish. Along the midside is a thick band of dark blotches that forms a fairly broad stripe. This is most noticeable in young specimens and may be indistinct in larger, older fish.

Similar Species. This is the only bass in the Okefenokee Swamp. It is readily distinguished from bream because of the deep notch between the spinous and soft dorsal rays and the elongate body that is not much compressed.

Distribution. Originally the largemouth bass ranged throughout much of central and southeastern United States west to northeastern Mexico. Apparently it did not occur east of the Appalachians from North Carolina through New England. Today, because of massive introductions, it ranges throughout almost all of the United States, extreme southern Canada, and Central America.

Habitat. The largemouth bass occurs in a variety of habitats but is generally restricted to slow-moving, quiet and standing waters. In the Okefenokee Swamp it is most commonly found in lakes, prairies, canals, sloughs, and the quiet backwater pools of larger streams and rivers. During the day individual specimens remain in fairly deep water in and around vegetation or submerged debris. They move into shallow water during evening to feed. Although adult fish are solitary, at times they will congregate in large numbers below dams, spillways, or the downstream side of sandbars in streams.

Biology. Because of its importance as a game fish, the largemouth bass has been studied extensively by fish biologists (28). Young largemouth bass feed on small crustaceans. As they increase in size, they shift to eating aquatic insects and larvae. Adult fish are voracious, feeding on smaller fishes, crayfish, and amphibians, as well as on nearly any other small animal that occurs in the water.

Spawning may take place as early as January and last until June. Nests are constructed by the males. A female may spawn with several males on different nests. The egg number—as many as 100,000—varies with the size of the females. The adhesive eggs, which are scattered along the bottom of the nest, hatch in 3–5 days. The male may remain with the young, which school for several weeks. Largemouth bass grow very rapidly, measuring 5–6 or more inches the first year and adding as many as 2–3 inches in succeeding years. They reach sexual maturity at 2–3 years. Longevity may be in excess of 10 years.

Comments. Throughout the South, as elsewhere, the largemouth bass is probably the most popular game fish, primarily because of its sporting qualities and its excellent flavor.

The largemouth bass inhabiting the Okefenokee Swamp are intergrades between the more northerly distributed *M. s. salmoides* and *M. s. floridanus*, endemic to peninsular Florida (3).

99

Black Crappie

Pomoxis nigromaculatus (Lesueur)

Other Local Names.   Crappy, calico bass, speckled perch, white perch.

Scientific Name.   *Pomoxis*, from two Greek words meaning "sharp opercle"; *nigromaculatus*, from two Latin words meaning "black-spotted."

Description.   The black crappie is larger than most sunfishes, ranging from 5 to 12 inches; however, it may exceed 14 inches. It has a very deep, slab-sided body with the greatest depth at the origin of the dorsal fin. The back is strongly arched, and the top of the head above the eye is markedly concave. The mouth is large, extending to the middle of the eye. The operculum is bony to the posterior edge, which is pointed. There are usually 7–8 spinous and 14–16 soft dorsal rays with no notch between. The caudal fin is markedly notched. The anal fin is nearly as long as the dorsal and has 6–7 spinous and 16–18 soft rays.

The top of the head, back, and upper sides are dark olive with a metallic blue or green cast. The lower sides and belly are silvery white. The back and sides are irregularly mottled with dark green or black blotches. The dorsal, caudal, and anal fins are heavily mottled with an irregular pattern of dark greenish or black spots. During breeding the backs of males become particularly dark.

Similar Species.   The black crappie can easily be distinguished from most other sunfishes because it has more than 3 anal spines and less than 9 dorsal spines. It might be confused with the flier, but the flier has 11–13 spinous dorsal rays.

Distribution.   This species has been widely introduced throughout most of the United States, excluding the Rocky Mountains. They are widely distributed in the Southeast and are common in both the St. Marys and Suwannee rivers.

Habitat.   The black crappie prefers warm, clean, quiet waters with sandy or muddy bottoms. It is not usually tolerant of turbid or silty waters. Individual fish congregate in relatively large schools and are almost always found near abundant vegetation or submerged debris. In the Okefenokee they are distributed primarily in the larger lakes, canals, and deep pools and backwaters of rivers and their larger tributary streams.

Biology.   Younger, smaller (up to about 6–7 inches) black crappie feed almost exclusively on crustaceans. Adult fish feed on a wide variety of smaller fishes and the larvae of aquatic insects (45, 73). They usually feed in fairly open water in the early morning or at night.

The breeding season extends from January to as late as June in the extreme Southeast. The male typically clears a small, shallow depression in or around aquatic vegetation for a nest. As many as 65,000 adhesive eggs are shed by the female. A single female may spawn with several males in different nests. The eggs are guarded and fanned (to provide aeration) by the male. The eggs hatch in 3–5 days, and the young continue to be guarded by the male until they disperse. Black crappie grow rapidly, as much as 4 inches the first year, and usually become sexually mature in the second year. Their maximum life span is 8–10 years.

Comments.   Because the black crappie grows to a fairly large size and has an excellent flavor, this species is a popular game fish. They are generally fished in relatively deep water in or around debris and/or vegetation. The best fishing is in the spring, when large numbers congregate in spawning areas.

Although we have seen no evidence of hybridization in the Okefenokee Swamp, black crappie are known to interbreed elsewhere, particularly with largemouth bass, warmouth, and bluegill (2).

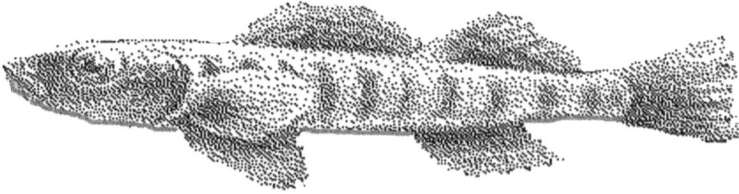

Swamp Darter

Etheostoma fusiforme (Girard)

Other Local Names.   None.

Scientific Name.   *Etheostoma,* the origin of this word is not evident; *fusiforme,* from the Latin for fusiform or spindle-shaped.

Description.   The swamp darter is small, reaching a maximum size of only 2 inches. It is elongate and slender in shape. The snout is somewhat rounded and the mouth is terminal. Two dorsal fins are present, the first of which has 9–11 spiny rays and the second, 10–12 soft rays. The anal fin has two spinous rays and 7–8 soft rays. The pelvic and anal rays, and sometimes the anal spines of breeding males, are covered by breeding tubercles. Genital papillae in females may be very elongate.

Coloration in the species may be extremely varied. Typically, the back and sides are dark brown and the belly, pale. The belly may be immaculate or have scattered spots. There are often 8–13 brown-to-black blotches along the sides below the lateral line. These blotches may or may not fuse into a lateral band and in some specimens may be undetectable. Specimens from the Okefenokee Swamp usually have 4 dark spots at the base of the caudal fin. In females the soft dorsal and anal fins are normally barred or spotted with dark pigment. In breeding males a median band of dark pigment is present on the spinous dorsal fin.

Similar Species. The swamp darter might be confused with small specimens of the blackbanded darter. The swamp darter, however, has a more definite, uniform brownish coloration and lacks the vertical bars seen even in juvenile blackbanded darters. Also, the blackbanded darter appears more robust than a similar-sized swamp darter.

Distribution. The species occurs from Maine south along the Atlantic Slope to Florida and west along the Gulf Coast to Texas. It also ranges up the Mississippi River to Tennessee and has been introduced in ponds in western North Carolina.

Habitat. The swamp darter is a bottom dweller and inhabits waters with little or no current. It almost always prefers sandy or muddy bottoms and lives near or among underwater debris such as vegetation, logs, and even discarded cans and bottles. In the Okefenokee Swamp it is found along the margins of lakes, usually in vegetation, and along the margins of streams, boat trails, and prairies. It sometimes is found in fairly deep waters.

Biology. The swamp darter is carnivorous and feeds on small crustaceans and aquatic insect larvae found in bottom substrates and vegetation (19).

The peak of the spawning season probably occurs from March to May, but breeding specimens have been found as late as October in the Suwannee River (19). Further south the species may breed throughout the year (55). The swamp darter does not appear to be territorial, and spawning acts consist of males approaching females from the rear, mounting, and beating her with tuberculate pelvic fins. Eggs are laid singly on vegetation (19). Growth and longevity studies on the swamp darter are not available, but published evidence indicates that they grow to about 1 inch in 2 months and that most probably live for about a year (19).

Comments. Two nominal subspecies of the swamp darter are recognized (19). *Etheostoma f. fusiforme* occurs from North Carolina northward, and *E. f. barratti* occurs from South Carolina southward throughout the rest of the range, including the Okefenokee Swamp.

The swamp darter is an excellent aquarium species, feeds quite well on live food such as the water flea *Daphnia,* and will spawn readily on suitable vegetation in captivity.

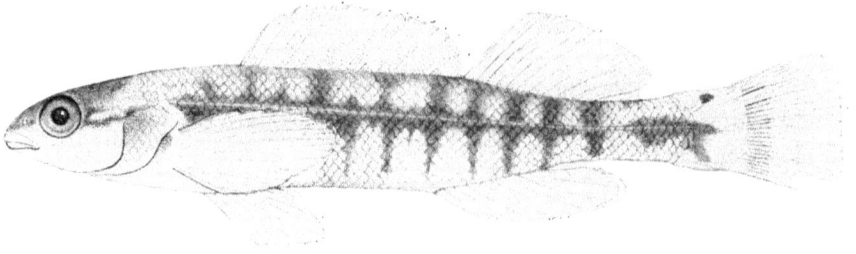

Blackbanded Darter

Percina nigrofasciata (Agassiz)

Other Local Names. Crawl-a-bottom.

Scientific Name. *Percina,* from the Greek meaning "little perch"; *nigrofasciata,* from two Latin words for "black" and "lined."

Description. This is a rather large darter that grows to a maximum length of over 3½ inches. The body shape is "perchlike"—rather elongate and robust. The body is scaled except for the breast, which is unscaled. The snout is somewhat pointed, and the moderately sized mouth is tilted downward. The black-banded darter has two separate dorsal fins; the first is spinous and the second, soft-rayed. The caudal fin is slightly emarginate.

The ground color of the fish is typically dark olive above and cream below. The species is characterized by a series of dark vertical bars superimposed on a dark lateral band that extends from the snout through the eye to the three spots at the base of the caudal fin (the lower two are usually fused). The sexes may be distinguished during the breeding season. Females normally have blotches below the lateral line in the spaces between the vertical bars. These blotches may also be evident during the nonbreeding season. The dorsal and anal fins may have dark spots. Males lack blotches and spots and are iridescent green.

Similar Species. Although very small specimens might be confused with the swamp darter, *Etheostoma fusiforme,* there are no similar species inhabiting the Okefenokee. In general the vertical bars along the sides are very distinctive even in small juvenile blackbanded darters. This characteristic is completely absent in the swamp darter.

Distribution.   The blackbanded darter occurs in Gulf Coast and Atlantic Slope drainages from the Lake Pontchartrain drainage in Louisiana eastward through Mississippi, Alabama, and Georgia to the Edisto River system in South Carolina, and as far south as the Suwannee and St. Johns river systems in Florida. The species is conspicuously absent from the St. Marys and Satilla river systems in Georgia. The blackbanded darter occurs only in the Suwannee River in the Okefenokee Swamp watershed.

Habitat.   Throughout its range, the blackbanded darter inhabits flowing waters, where current velocities may range from slow to torrential, and generally prefers some form of substrate, such as gravel, snags, or aquatic macrophytes (84). The fish rarely is found in waters with silty or muddy bottoms. The species is not recorded from the refuge boundaries, having been collected only in the Suwannee River.

Biology.   This species is a carnivorous, visual feeder. The bulk of its diet is composed of small crustaceans and aquatic insect larvae. It feeds primarily in the morning and late afternoon (58).

In the extreme southern portions of its range the blackbanded darter breeds throughout the year (55), although the most active breeding is in spring (59). The number of eggs released by the female may range between 38 and 250, depending upon her size.

Comments.   The blackbanded darter may be easily observed resting on the sandy bottom in shallow water where some current is evident. It is a difficult fish to collect and, when approached, will often dart rapidly forward, from side to side, for short distances seeking cover. In its preferred habitat, especially during periods of low water, literally dozens may be seen on the downcurrent side of rocks, logs, or vegetation by the careful observer.

Glossary

Abdomen. Ventral part of the trunk from the pectoral fins to the anus.

Adipose fin. A dorsally located, unpaired, fleshy-lobed median fin located behind the dorsal fin; occurs in catfishes.

Anal fin. The unpaired median fin on the ventral surface behind the anus.

Annulae. A series of growth rings present on scales.

Anterior. Referring to the front.

Barbel. A worm-shaped, fleshy, tactile organ projecting from the lower portion of the head; usually associated with catfish.

Belly. Ventral part of the trunk from the pectoral fins to the anus.

Body depth. The greatest vertical distance from the dorsal part of the body to the ventral part of the body, excluding fins.

Branchiostegal rays. A series of elongate bones that support the gill membranes. They are arranged fanwise at the ventral edges of the gill covers.

Breast. Ventral part of the trunk anterior to the pectoral fins.

Carnivorous. Feeding on animal tissue.

Catadromous. Fishes, like eels, which spend most of their lives in fresh water but migrate to the sea to reproduce.

Caudal fin. The tail fin.

Caudal peduncle. The constricted portion of the tail between the posterior end of the anal fin and the base of the caudal fin.

Compressed. Flattened from the sides.

Crustaceans. A group of aquatic organisms with a hard outer shell; includes water fleas, crayfish, etc.

Ctenoid. Scale type characterized by a toothed or spiny margin.

Cycloid. Scale type characterized by a smooth, free margin.

Dentary. The most prominent of the lower jawbones; usually bears teeth.

Detritus. Decomposing organic matter.

Dimorphism. Having two forms; males and females are structurally different.

Diurnal. Referring to the daytime.

Dorsal. Referring to the top or back.

Ecology. The study of the interrelationships of organisms and their natural environment.

Emarginate. A notched or concave margin, but not so deeply notched as to be considered forked.

Fin ray. Elements that support the fin.

Frenum. Fleshy bridge between the upper lip and snout; the presence of this connecting membrane restricts protraction of the premaxillaries of the upper jaw.

Ganoid. Thick bony scales that do not overlap and are rhomboid or diamond-shaped.

Gill. The respiratory organ in fish consisting of vascularized gill filaments that extract oxygen from the water.

Gonopodium. Modified anal fin in some male fish used as an accessory copulatory organ.

Gravid. A condition in which the female is swollen with eggs or in which the male is greatly distended with sperm.

Gular plate. A bony plate that lies between the lower jawbones in primitive bony fish.

Habitat. The normal environment where an organism lives.

Head. Region of the body extending from the tip of the snout to the posterior margin of the operculum.

Head length. The distance from the anteriormost part of the snout to the posterior edge of the operculum.

Herbivorous. Feeding on plant material.

Heterocercal. Tail shape in which the axis of the vertebral column extends out into the upper lobe of the caudal fin.

Homocercal. Tail shape in which the vertebral column ends at the base of the caudal fin and the fin itself is symmetrical top and bottom.

Hybrid. The result of crossing or reproduction between individuals of separate species.

Ichthyology. The study of fishes.

Interopercle. The bone ventral and anterior to the subopercle.

Invertebrate. Animals that lack a vertebral column or backbone.

Isthmus. The fleshy region of the throat between the gills.

Larva. Early developmental stage in a fish between hatching and transformation into a juvenile.

Lateral. The sides or toward the sides.

Lateral line. A series of pores located in scales that connect with an underlying canal; typically extends from the back of the head to the base of the caudal fin along the side of the body.

Leptocephalus. The transparent, almost ribbonlike larval form of the eel.

Littoral. Relating to the edge or near edge of a body of water.

Maxillary. The bone behind (and sometimes above) the premaxillary; together they make up the upper jaw.

Medial. The midline or toward the midline.

Molluscs. A group of aquatic organisms including clams, oysters, snails, etc.

Morphology. Relating to shape, form, or structure.

Naked. Absence of scales.

Nocturnal. Referring to night.

Ocellus. An eyelike spot of color.

Omnivorous. Feeding on both animal and vegetable substances.

Opercle. The largest and usually most dorsal and posterior bone of the gill cover (operculum).

Operculum. The gill cover, including all the bones that make up the flap that covers the gills.

Orbit. The bony eye socket.

Oviparous. Egg laying in which the eggs develop and hatch outside the female's body.

Palatine. Bones lateral to the vomer in the roof of the mouth; usually bear teeth (palatine teeth).

Pectoral fin. Frontmost of the paired fins, located directly behind the head; correspond to the forelimbs or arms of higher vertebrates.

Pelvic fin. Hindmost of the paired fins; usually located midtrunk but may be directly behind and below the pectoral fins.

Phylogenetic. Referring to evolutionary history.

Physoclistus. A condition in which there is no duct connecting the swim bladder to the pharynx.

Physostomous. A condition in which the swim bladder is connected to the pharynx by way of a duct permitting the use of the swim bladder as an accessory respiratory device.

Piscivorous. Feeding on fish.

Plankton. Weakly swimming or passively floating, small aquatic plants and animals.

Posterior. Referring to the hind or rear.

Premaxillary. The anteriormost bone in the upper jaw; usually bears teeth.

Preopercle. The bone anterior to the opercle and dorsal to the interopercle; it makes up the anterior edge of the gill cover; its posterior border may be free and may be serrated.

Protractile. Extension or projection of the upper jaw forward from the snout in those fishes not having the upper jaw connected to the snout by a fleshy bridge (the frenum).

Ray. Fin support.

Scale count. The number of scales counted on various parts of the body.

Serration. Having a toothed or sawlike edge.

Snout. The region of the head anterior to the orbit.

Snout length. The distance from the anteriormost part of the snout to the anterior border of the orbit.

Species. A group of interbreeding or potentially interbreeding individuals that share a common inheritance.

Spinous ray. Hard, sharply pointed fin ray; not branched or segmented.

Soft ray. Soft fin ray; segmented and often branched.

Subopercle. The bone posterior and ventral to the opercle.

Subspecies. Local populations of a species that inhabit a geographic subdivision of the range of the species and differ from other populations.

Swim bladder. A thin-walled, membranous, gas-filled sac lying above the viscera in the abdomen; functions as a hydrostatic organ but may be used as an accessory respiratory organ.

109

Tail. Region from the anus to the tip of the caudal fin.

Taxonomy. The study of the kinds and relationships of various groups of organisms.

Thorax. Ventral part of the trunk anterior to the pectoral fins.

Total length. The distance from the tip of the snout to the tip of the caudal fin.

Trunk. Region of the body extending from the operculum to the anus or vent.

Ventral. Referring to the bottom or belly.

Viviparous. Giving birth to live young as opposed to laying eggs.

Vomer. A median bone in the roof of the mouth that usually bears teeth (vomerine teeth).

References

1. Arndt, Rudolf. 1971. Ecology and behavior of the cyprinodont fishes *Adinia xenica, Lucania parva, Lucania goodei,* and *Leptolucania ommata.* Ph.D. dissertation, Cornell University, Ithaca, N.Y. 344 pp.
2. Avise, J. C., D. O. Straney, and M. H. Smith. 1977. Biochemical genetics of sunfish. IV. Relationships of centrarchid genera. *Copeia* 1977:250–58.
3. Bailey, R. M., and C. L. Hubbs. 1949. The black basses (*Micropterus*) of Florida, with description of a new subspecies. *University of Michigan Museum of Zoology Occasional Papers* 516:1–40.
4. Bailey, R. M., H. E. Winn, and C. L. Smith. 1954. Fishes from the Escambia River, Alabama and Florida, with ecologic and taxonomic notes. *Proceedings of the Academy of Natural Sciences of Philadelphia* 106:109–64.
5. Barnickol, P. G. 1941. Food habits of *Gambusia affinis* from Reelfoot Lake, Tennessee, with special reference to malarial control. *Report of the Reelfoot Lake Biological Station* 5:5–13.
6. Beck, K. C., J. H. Reuter, and E. M. Perdue. 1974. Organic and inorganic geochemistry of some coastal plain rivers of the southeastern United States. *Geochimica et Cosmochimica Acta* 44:341–64.
7. Benke, A. C., D. M. Gillespie, and F. K. Parrish. 1979. Biological basis for assessing impacts of channel modification: Invertebrate production, drift, and fish feeding in a southeastern blackwater river. Environmental Research Center, Georgia Institute of Technology, Atlanta, 06–79. 187 pp.
8. Berry, F. H. 1955. Food of the mudfish (*Amia calva*) in Lake Newnan, Florida, in relation to its management. *Quarterly Journal of the Florida Academy of Sciences* 18: 69–75.
9. Blood, E. R. 1982. Surface water hydrology and biogeochemistry of the Okefenokee Swamp. Ph.D. dissertation, University of Georgia, Athens. 194 pp.
10. Böhlke, J. 1956. A new pygmy sunfish from southern Georgia. *Notulae Naturae (Philadelphia)* 294:1–11.
11. Breder, C. M., Jr. 1935. The reproductive habits of the common catfish, *Ameiurus nebulosus* (Lesueur), with a discussion of their significance in ontogeny and phylogeny. *Zoologica* 19:173–85.
12. Breder, C. M., Jr. 1939. Variations in the nesting habits of *Ameiurus nebulosus* (Lesueur). *Zoologica* 24:367–68.

13. Breder, C. M., Jr., and D. E. Rosen. 1966. *Modes of reproduction in fishes.* Neptune City, New Jersey: T. F. H. Publications. 941 pp.

14. Brown, J. L. 1958. Geographic variation in southeastern populations of the cyprinodont fish *Fundulus notti* (Agassiz). *American Midland Naturalist* 59(2):877–78.

15. Cahn, A. R. 1927. An ecological study of southern Wisconsin fishes, the brook silverside (*Labidesthes sicculus*) and the cisco (*Leuchicthys artedi*) in relation to the region. *Illinois Biological Monographs* 11:1–151.

16. Carr, M. H. 1946. The breeding habits of the eastern stumpknocker, *Lepomis punctatus*. *Quarterly Journal of the Florida Academy of Sciences* 9:101–6.

17. Clark, K. E. 1977. Reproductive biology of *Noturus leptacanthus* Jordan. A.S.I.H. Abstracts of the 57th Annual Meeting.

18. Cohen, A. D., D. J. Cassagrande, M. J. Andrejko, and G. R. Best. 1984. *The Okefenokee Swamp: Its natural history, geology, and geochemistry.* Los Alamos, New Mexico: Wetland Surveys. 709 pp.

19. Collette, B. B. 1962. The swamp darters of the subgenus *Hololepis* (Pisces, Percidae). *Tulane Studies in Zoology* 9:115–211.

20. Conley, J. M. 1966. Ecology of the flier, *Centrarchus macropterus* (Lacepède), in southeast Missouri. M.A. thesis, University of Missouri, Columbia. 119 pp.

21. Crossman, E. J. 1966. A taxonomic study of *Esox americanus* and its subspecies in eastern North America. *Copeia* 1966:1–20.

22. Crossman, E. J., and K. Buss. 1965. Hybridization in the family Esocidae. *Journal of the Fisheries Research Board of Canada* 22:1261–92.

23. Dahlberg, M. D., and D. C. Scott. 1966. The freshwater fishes of Georgia. *Bulletin of the Georgia Academy of Science* 29:1–64.

24. Doering, J. A. 1960. Quarternary surface formations of the southern part of the Atlantic Coastal Plain. *Journal of Geology* 68:182–202.

25. Eales, J. G. 1968. The eel fisheries of eastern Canada. *Bulletin of the Fisheries Research Board of Canada,* p. 166. 79 pp.

26. Eddy, S. F., and T. Surber. 1943. *Northern fishes.* Minneapolis: University of Minnesota Press. 276 pp.

27. Emmig, J. W. 1966a. Bluegill sunfish. In *Inland fisheries management,* ed. A. Calhoun, pp. 325–92. Sacramento: California Fish and Game Department. 546 pp.

28. Emmig, J. W. 1966b. Largemouth bass. In *Inland fisheries management,* ed. A. Calhoun, pp. 332–53. Sacramento: California Fish and Game Department. 546 pp.

29. Fahay, M. P. 1978. Biological and fisheries data on the American eel, *Anguilla rostrata* (Lesueur). Sandy Hook Laboratory, Highlands, New Jersey, Technical Services Report 17. 422 pp.

30. Forbes, S. A., and R. E. Richardson. 1920. *The Fishes of Illinois.* 2d ed. Illinois Natural History Survey 3. 357 pp.

31. Foster, N. R. 1967. Comparative studies in the evolution of reproductive behavior in killifishes. *Studies in Tropical Oceanography, Institute of Marine Science, University of Miami* 5:549–66.

32. Freeman, B. J., and M. C. Freeman. 1985. Production of fishes in a subtropical blackwater ecosystem: The Okefenokee Swamp. *Limnology and Oceanography* 30:686–92.
33. Gassaway, R. D., Jr. 1976. Factors associated with catch of fishes in the Lower Coastal Plain tributary streams. M.S. thesis, Mississippi State University. 79 pp.
34. German, J. F. 1973. Age and growth, reproduction and food habits of the warmouth in the Okefenokee Swamp and the Suwannee River. In *Annual progress reports*. Statewide fisheries investigations. Vol. 2, pp. 55–70. Atlanta: Georgia Department of Natural Resources.
35. Harper, F., and D. E. Presley. 1981. *Okefinokee Album*. Athens: University of Georgia Press. 194 pp.
36. Hellier, T. R., Jr. 1967. The fishes of the Santa Fe River system. *Bulletin of the Florida State Museum* 2:1–46.
37. Holder, D. R. 1970. A study of fish movements from the Okefenokee Swamp into the Suwannee River. *Proceedings of the S.E. Association of the Game and Fish Commission* 18:247–55.
38. Holder, D. R. 1973a. Age and growth, reproduction and food habits of bowfin in the Okefenokee Swamp and in the Suwannee River. In *Annual progress reports*. Statewide fisheries investigations. Vol. 2, pp. 44–54. Atlanta: Georgia Department of Natural Resources.
39. Holder, D. R. 1973b. Stream creel census, Suwannee River and Okefenokee Swamp. In Annual progress reports. Statewide fisheries investigations. Vol. 3, pp. 64–83. Atlanta: Georgia Department of Natural Resources.
40. Holloway, A. D. 1954. Notes on the life history and management of the shortnose and longnose gars in Florida waters. *Journal of Wildlife Management* 18:433–49.
41. Hubbs, C. L. 1930. Materials for a revision of the catostomid fishes of eastern North America. *Miscellanous Publications of the Museum of Zoology, University of Michigan* 20:1–47.
42. Hubbs, C. L., and K. F. Lagler, 1964. *Fishes of the Great Lakes Region*. Ann Arbor: University of Michigan Press. 213 pp.
43. Hunt, B. P. 1953. Food relationship between spotted gar and other organisms in the Tamiami Canal, Dade County, Florida. *Transactions of the American Fisheries Society* 82:13–33.
44. Jackson, S. W., Jr. 1957. Comparison of the age and growth of four fishes from the Lower and Upper Spavinaw Lakes, Oklahoma. *Proceedings of the 11th Annual Conference of the S.E. Association of the Game and Fish Commission*, pp. 232–49.
45. Keast, A. 1968. Feeding biology of the black crappie, *Pomoxis nigromaculatus*. *Journal of the Fisheries Research Board of Canada* 25:285–97.
46. Keast, A., and D. Webb. 1966. Mouth and body form relative to feeding ecology in the small fish fauna of a small lake, Lake Opinicon, Ontario. *Journal of the Fisheries Research Board of Canada* 23:1845–67.
47. Krumholz, L. A. 1948. Reproduction of the western mosquitofish, *Gambusia a.*

113

affinis (Baird and Girard), and its use in mosquito control. *Ecological Monographs* 18:1–43.

48. Laerm, J., B. J. Freeman, L. J. Vitt, J. M. Myers, and L. E. Logan. 1980. Vertebrates of the Okefenokee Swamp. *Brimleyana* 4:47–73.

49. Lagler, K. F., and F. V. Hubbs. 1940. Food of the long-nosed gar (*Lepisosteus osseus oxyurus*) and the bowfin (*Amia calva*) in southern Michigan. *Copeia* 1940:239–41.

50. Larimore, R. W. 1957. Ecological life history of the warmouth (Centrarchidae). *Illinois Natural History Survey Bulletin* 27:1–83.

51. Lee, R. S., C. R. Gilbert, C. H. Hocutt, R. E. Jenkins, D. E. McAllister, and J. R. Stauffer, Jr. 1980. Atlas of North American freshwater fishes. North Carolina Biological Survey 12. 867 pp.

52. Leitholf, E. 1917. *Fundulus chrysotus. Aquatic Life* 2(11):141–42.

53. Lowe, J. A., and S. B. Jones, Jr. 1984. Checklist of vascular plants of the Okefenokee Swamp. In *The Okefenokee Swamp: Its natural history, geology, and geochemistry*, ed. A. D. Cohen, D. J. Cassagrande, M. J. Andrejko, and G. R. Best, pp. 702–9. Los Alamos, New Mexico: Wetland Surveys.

54. McCabe, B. C. 1958. Esox niger *Lesueur. Tabular treatment of the life history and ecology of the chain pickerel*. National Academy of Sciences Committee on the Handbook of Biological Data. 45 pp.

55. McLane, W. M. 1955. The fishes of the St. Johns River system. Ph.D. dissertation, University of Florida, Gainesville. 361 pp.

56. McSwain, L. E., and R. M. Jennings. 1972. Spawning behavior of the sucker, *Minytrema melanops* (Rafinesque). *Transactions of the American Fisheries Society* 101:738–40.

57. Mansuetti, R., and H. J. Elser. 1953. Biology, age and growth of the mud sunfish, *Acantharchus pomotis*, in Maryland. *Copeia* 1953:117–19.

58. Mathur, D. 1973a. Some aspects of life history of the blackbanded darter, *Percina nigrofasciata* (Agassiz) in Halawakee Creek, Alabama. *American Midland Naturalist* 89:381–93.

59. Mathur, D. 1973b. Food habits and feeding chronology of the blackbanded darter, *Percina nigrofasciata* (Agassiz) in Halawakee Creek, Alabama. *Transactions of the American Fisheries Society* 102:48–55.

60. Miller, E. E. 1966. Channel catfish. In *Inland fisheries management*, pp. 440–63. Sacramento: California Fish and Game Department.

61. Ming, A. D. 1968. Life history of the grass pickerel, *Esox americanus vermiculus* in Oklahoma. Oklahoma Fisheries Research Laboratory, Bulletin 8. 66 pp.

62. Moody, H. L. 1957. A fisheries sutdy of Lake Panasoffkee, Florida. *Quarterly Journal of the Florida Academy of Sciences* 20:21–88.

63. Nelson, J. S. 1968. Life history of the brook silverside, *Labidesthes sicculus*, in Crooked Lake, Indiana. *Transactions of the American Fisheries Society* 97:293–96.

64. Palmer, E. L., and A. H. Wright. 1920. A biological reconnaissance of the Okefenokee Swamp in Georgia: The fishes. *Journal of the Iowa Academy of Science* 27:353–77.

65. Parrish, F. R., and E. J. Rykiel, Jr. 1979. Okefenokee Swamp origin: Review and reconsideration. *Journal of the Elisha Mitchell Science Society* 95:17–31.
66. Peckham, R. S., and C. F. Dineen. 1957. Ecology of the central mudminnow, *Umbra limi* (Kirtland). *American Midland Naturalist* 58:222–31.
67. Pflieger, W. L. 1975. *The fishes of Missouri.* Missouri Department of Conservation. 343 pp.
68. Phillips, C. 1958. An unusually colored garfish, *Lepisosteus platyrhincus. Copeia* 1958:331.
69. Price, G. 1915. The black-banded sunfish. *Aquatic Life* 1:45–46.
70. Raney, E. C., and D. A. Wester. 1940. The food and growth of the young of the common bullhead, *Ameiurus nebulosus* (Lesueur) in Cayuga Lake, New York. *Transactions of the American Fisheries Society* 69:205–9.
71. Reed, H. D. 1907. The poison glands of *Noturus* and *Schilbeodes. American Naturalist* 41:553–66.
72. Reeves, J. D., and G. A. Moore. 1949. *Lepomis marginatus* (Holbrook) in Oklahoma. *Proceedings of the Oklahoma Academy of Science* 30:41–42.
73. Reid, G. K. 1950. Food of the black crappie, *Pomoxis nigromaculatus* (Lesueur), in Orange Lake, Florida. *Transactions of the American Fisheries Society* 79:145–54.
74. Rivas, L. R. 1966. The taxonomic status of the cyprinodontid fishes *Fundulus notti* and *F. lineolatus. Copeia* 1966:353–54.
75. Robins, C. R., R. M. Balley, C. E. Bond, J. R. Brooker, E. A. Lachner, R. N. Lea, and W. B. Scott. 1980. *A list of common and scientific names of fishes from the United States and Canada.* 4th ed. American Fisheries Society Special Publication 12. 174 pp.
76. Rykiel, E. J. 1977. The Okefenokee Swamp watershed: Water balance and nutrient budgets. Ph.D. dissertation, University of Georgia, Athens. 246 pp.
77. Rykiel, E. J. 1984. General hydrology and mineral budgets for Okefenokee Swamp: Ecological significance. In *The Okefenokee Swamp: Its natural history, geology, and geochemistry,* ed. R. D. Cohen, D. J. Cassagrande, M. I. Andrejko, and G. R. Best, pp. 212–28. Los Alamos, New Mexico: Wetland Surveys.
78. Schmidt, J. 1922. The breeding places of the eel. *Philadelphia Transactions of the Royal Society of London,* series B 221:179–208.
79. Schwartz, F. J. 1961. Food, age, growth, and morphology of the blackbanded sunfish, *Enneacanthus c. chaetodon,* in Smithville Pond, Maryland. *Chesapeake Science* 2:82–88.
80. Scott, W. B. 1938. The food of *Amia* and *Lepisosteus. Investigations of Indiana Lakes and Streams* 1:110–15.
81. Shuffman, R. J. 1955. Age and rate of growth of the yellow bullhead in Reelfoot Lake, Tennessee. *Journal of the Tennessee Academy of Science* 30:4–7.
82. Short, L. 1956. A new pygmy sunfish. *Aquarium (Philadelphia)* 25(4):133–35.
83. Snow, H., A. Enign, and J. Klingbiel. 1960. The Bluegill: Its life history, ecology, and management. Wisconsin Conservation Department Publication 230:1–14.

84. Sutkus, R. D., and J. S. Ramsey. 1967. *Percina aurolineata*, a new percid fish from the Alabama river system and a discussion of ecology, distribution, and hybridization of darters of the subgenus *Hadropterus*. *Tulane Studies in Zoology* 13:129–45.

85. Sweeney, E. F. 1972. The systematics and distribution of the centrarchid tribe Enneacanthini. Ph.D. dissertation, Boston University.

86. Turner, C. L. 1937. Reproductive cycles and superfetation in peociliid fishes. *Biological Bulletin (Woods Hole)* 72:145–64.

87. Vladykov, V. D. 1964. Quest for the true breeding areas of the American eel (*A. rostrata* Lesueur). *Journal of the Fisheries Research Board of Canada* 21:1523–30.

88. White, D. S., and K. H. Haag. 1977. Foods and feeding habits of the spotted sucker, *Minytrema melanops* (Rafinesque). *American Midland Naturalist* 98:137–46.

89. Wich, K., and J. W. Mullen. 1958. A compendium of the life history and ecology of the chain pickerel, *Esox niger* Lesueur. Massachusetts Division of Fish and Game Fisheries Bulletin, p. 22. 27 pp.

90. Wiley, E. O. 1977. The phylogeny and systematics of the *Fundulus notti* species group (Teleostei: Cyprinodontidae). *Occasional Papers of the Museum of Natural History, University of Kansas* 66:1–31.

Index

Acantharchus pomotis, 80–81
American eel, 38–39
Amia calva, 36–37
Amiidae, 23. *See also* Bowfin, 36–37
Anatomy. *See* Fish: anatomy, 4–9
Anguilla rostrata, 38–39
Anguillidae, 23–24. *See also* American eel, 38–39
Aphredoderidae, 26. *See also* Pirate perch
Aphredoderus sayanus, 60–61
Atherinidae, 27. *See also* Brook silverside

Banded sunfish, 88–89
Banded topminnow, 64–65
Blackbanded darter, 104–5
Blackbanded sunfish, 84–85
Black crappie, 100–101
Bluegill, 92–93
Bluespotted sunfish, 86–87
Bowfin, 36–37
Bowfins. *See* Amiidae
Brook silverside, 74–75
Brown bullhead, 52–53

Catostomidae, 25. *See also* Lake chubsucker, Spotted sucker
Catfishes. *See* Ictaluridae
Centrarchidae, 28–30. *See also* Banded sunfish, Blackbanded sunfish, Black crappie, Bluegill, Bluespotted sunfish, Dollar sunfish, Flier, Largemouth bass, Mud sunfish, Spotted sunfish, Warmouth
Centrarchus macropterus, 82–83

Chain pickerel, 42–43
Channel catfish, 54–55
Common names, use of, 1
Cyprinodontidae, 26–27. *See also* Banded topminnow, Golden topminnow, Lined topminnow, Pygmy killifish

Darters. *See* Percidae
Dollar sunfish, 94–95
Drought, 12–13

Eastern mudminnow, 44–45
Eels. *See* Anguillidae
Elassoma evergladei, 76–77
Elassoma okefenokee, 78–79
Elassomidae, 28. *See also* Everglades pygmy sunfish, Okefenokee pygmy sunfish
Enneacanthus chaetodon, 84–85
Enneacanthus gloriosus, 86–87
Enneacanthus obesus, 88–89
Erimyzon sucetta, 46–47
Esocidae, 24. *See also* Chain pickerel, Redfin pickerel
Esox americanus, 40–41
Esox niger, 42–43
Etheostoma fusiforme, 102–3
Everglades pygmy sunfish, 76–77

Fires, 12–13
Fish: anatomy, 4–9; definition of, 4
Fishes: definition of, 4
Flier, 82–83
Florida gar, 34–35

Fundulus chrysotus, 62–63
Fundulus cingulatus, 64–65
Fundulus lineolatus, 66–67

Gambusia affinis, 70–71
Gars. *See* Lepisosteidae
Golden topminnow, 62–63

Heterandria formosa, 72–73

Ictaluridae, 25–26. *See also* Brown
  bullhead, Channel catfish, Speckled
  madtom, Tadpole madtom, Yellow
  bullhead
Ictalurus natalis, 50–51
Ictalurus nebulosus, 52–53
Ictalurus punctatus, 54–55

Keys, use of, 19–20
Keys: to families, 21–22; to species,
  23–30
Killifishes. *See* Cyprinodontidae

Labidesthes sicculus, 74–75
Lake chubsucker, 46–47
Largemouth bass, 98–99
Least killifish, 72–73
Lepisosteidae, 23. *See also* Florida gar
Lepisosteus platyrhincus, 34–35
Lepomis gulosus, 90–91
Lepomis machrochirus, 92–93
Lepomis marginatus, 94–95
Lepomis punctatus, 96–97
Leptolucania ommata, 68–69
Lined topminnow, 66–67
Livebearers. *See* Poeciliidae

Micropterus salmoides, 98–99
Minytrema melanops, 48–49
Mosquitofish, 70–71
Mudminnows. *See* Umbridae
Mud sunfish, 80–81

Noturus gyrinus, 56–57
Noturus leptacanthus, 58–59

Okefenokee pygmy sunfish, 78–79
Okefenokee Swamp: climate of, 12–13;
  geology of, 13; history of ichthyology
  in, 16–17; hydrology of, 9–12; origin
  of, 14–15; vegetation of, 15–16; water
  chemistry of, 13–14

Perches. *See* Percidae
Percidae, 30. *See also* Blackbanded
  darter, Swamp darter
Percina nigrofasciata, 104–5
Pikes. *See* Esocidae
Pirate perch, 60–61
Pirate perches. *See* Aphredoderidae
Plants. *See* Okefenokee Swamp:
  vegetation of
Poeciliidae, 27. *See also* Least killifish,
  Mosquitofish
Pomoxis nigromaculatus, 100–101
Pygmy killifish, 68–69
Pygmy sunfishes. *See* Elassomidae

Redfin pickerel, 40–41
Regional fauna, comparison with, 18–19

Silversides. *See* Atherinidae
Species: definition of, 2
Speckled madtom, 58–59
Spotted sucker, 48–49
Spotted sunfish, 96–97
Sunfishes. *See* Centrarchidae
Swamp darter, 102–3

Tadpole madtom, 56–57
Taxa, description of, 1

Umbra pygmaea, 44–45
Umbridae, 24. *See also* Eastern
  mudminnow

Warmouth, 90–91

Yellow bullhead, 50–51

www.ingramcontent.com/pod-product-compliance
Lightning Source LLC
Chambersburg PA
CBHW020708270326

41928CB00005B/320